T0140006

Unsupervised and Semi-Supervised Learning

Series Editor
M. Emre Celebi, Computer Science Department, MCST 303, Conway, AR, USA

Springer's Unsupervised and Semi-Supervised Learning book series covers the latest theoretical and practical developments in unsupervised and semi-supervised learning. Titles – including monographs, contributed works, professional books, and textbooks – tackle various issues surrounding the proliferation of massive amounts of unlabeled data in many application domains and how unsupervised learning algorithms can automatically discover interesting and useful patterns in such data. The books discuss how these algorithms have found numerous applications including pattern recognition, market basket analysis, web mining, social network analysis, information retrieval, recommender systems, market research, intrusion detection, and fraud detection. Books also discuss semi-supervised algorithms, which can make use of both labeled and unlabeled data and can be useful in application domains where unlabeled data is abundant, yet it is possible to obtain a small amount of labeled data.

Topics of interest in include:

- Unsupervised/Semi-Supervised Discretization
- Unsupervised/Semi-Supervised Feature Extraction
- Unsupervised/Semi-Supervised Feature Selection
- Association Rule Learning
- Semi-Supervised Classification
- Semi-Supervised Regression
- Unsupervised/Semi-Supervised Clustering
- Unsupervised/Semi-Supervised Anomaly/Novelty/Outlier Detection
- Evaluation of Unsupervised/Semi-Supervised Learning Algorithms
- Applications of Unsupervised/Semi-Supervised Learning

While the series focuses on unsupervised and semi-supervised learning, outstanding contributions in the field of supervised learning will also be considered. The intended audience includes students, researchers, and practitioners.

More information about this series at http://www.springer.com/series/15892

Xiangtao Li • Ka-Chun Wong
Editors

Natural Computing for Unsupervised Learning

Editors
Xiangtao Li
Department of Information Sciences
and Technology
Northeast Normal University
Changchun, Jilin, China

Ka-Chun Wong
City University of Hong Kong
Kowloon Tong, Hong Kong

ISSN 2522-848X ISSN 2522-8498 (electronic)
Unsupervised and Semi-Supervised Learning
ISBN 978-3-030-07508-8 ISBN 978-3-319-98566-4 (eBook)
https://doi.org/10.1007/978-3-319-98566-4

This Springer imprint is published by the registered company Springer Nature Switzerland AG
The registered company address is: Gewerbestrasse 11, 6330 Cham, Switzerland

Contents

Part I
Advances in Natural Computing

Chapter 1
Detailed Modeling of CSC-STATCOM with Optimized PSO Based Controller

Sandeep Gupta

1.1 Introduction

The STATCOM can be classified by their inverter configurations. According to this, there are two different configurations: using a voltage source converter (VSC) or a current source converter (CSC). When STATCOMs are used with different configurations, CSC topology has distinct benefits over VSC topology, which is discussed in the earlier chapter. Therefore in brief, the direct output of CSC is a controllable ac current, whereas that of a VSC is a controllable ac voltage. When SPWM (Sinusoidal Pulse Width Modulation) technique [1] is used to control both converters, then the values of the harmonic contents are directly proportional to the magnitudes of the fundamental parts of their direct output quantities. In most electric power transmission systems with normal system conditions, the current produced by STATCOM is a small part of the system current. Therefore, when CSC topology is applied in the system, the current harmonic components are also small. But, when VSC topology is applied for a small reactive current component, the output voltage of VSC is large and near to the system voltage. These outcomes in large voltage harmonic components are proportional to current harmonic components that are bigger than those generated by CSC topology and thus higher expensive to filter. The dc side energy storage is also another point for comparison of these converters. For the reactive power compensation by the STATCOM, the value of dc-side energy storage device in CSC topology is lower than that in VSC topology.

The three-phase CSC-STATCOM is a high efficient shunt connected FACTS device. Almost sinusoidal three-phase balanced currents are produced by CSC-STATCOM with flexible magnitudes. These currents also lead or lag the related line

S. Gupta (✉)
JECRC University, Jaipur, Rajasthan, India

X. Li, K.-C. Wong (eds.), *Natural Computing for Unsupervised Learning*, Unsupervised and Semi-Supervised Learning,
https://doi.org/10.1007/978-3-319-98566-4_1

voltages by an angle of 90°. Thus, the three-phase CSC-STATCOM can be used in many applications such as Ward-Leonard drives, to control and compensation of rapidly fluctuating reactive power demands due to different types of loading.

In this chapter, first the system description and modeling of three phase current source converter (CSC) based STATCOM is presented. Nonlinearity of this modeling is removed with the help of power balance equation and dq-transformation. After that pole-shifting and LQR based controllers are designed for the CSC-STATCOM. But the best state-feedback gain matrices for different controllers are obtained laboriously through trial and error method, although time-consuming. So this problem is solved with the help of proposed particle swarm optimization (PSO) based AI technique method to search the best values of state-feedback gain matrices in a very short time. In this chapter, comparison in between different methods is also presented with simulation results. The feasibility of the proposed scheme is demonstrated through simulation in MATLAB.

1.2 System Configuration and Modeling of CSC-Based STATCOM

To verify the response of the CSC-STATCOM on dynamic performance, the mathematical modeling and control strategy of a CSC based STATCOM are presented. So in the design of controller for CSC based STATCOM, the state-space equations from the CSC based STATCOM circuit are introduced. To minimize the complexity of mathematical calculation, the concept of dq-transformation has been applied in the test system circuit, which makes the d and q components as independent parameters. Figure 1.1 shows the circuit diagram of a typical CSC based STATCOM [2–6]. In this figure, six power electronics-based switches are used. These switches have fully controllable characteristic with bipolar voltage blocking and unidirectional

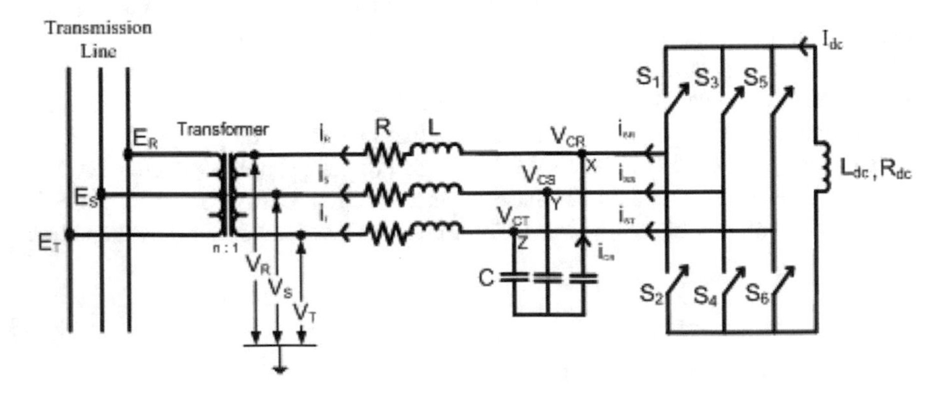

Fig. 1.1 Equivalent circuit of CSC based STATCOM

current-carrying abilities. In this figure, the CSC-STATCOM is connected to the transmission line through a shunt transformer.

CSC-STATCOM has dc-link reactor for the dc-side energy storage device. The dc-link reactor is the series combination of dc-link inductance (Ldc) and its internal resistance (Rdc). Internal resistance (Rdc) of the dc-link inductor also represents the converter switching and conduction losses. In Fig. 1.1, series R-L impedance is provided by an ideal transformer. Based on the equivalent circuit of CSC-STATCOM shown in Fig. 1.1, we can derive the mathematical model of CSC-STATCOM as follows:
where:

i_R, i_S, i_T	line currents;
E_R, E_S, E_T	line voltages;
V_R, V_S, V_T	transformer secondary side voltages;
V_{CR}, V_{CS}, V_{CT}	voltages across the filter capacitors;
i_{SR}, i_{SS}, i_{ST}	converter terminal currents;
I_{dc}	dc-side current;
R_{dc}	dc-link inductors internal resistance;
L_{dc}	smoothing inductor (dc-link);
C	filter capacitance;
L	line reactors inductance;
R	line reactors resistance

From circuit principles, converter ac side differential equation is represented by the Eq. (1.1) for phase R.

$$V_{CR} = Ri_R + L\frac{di_R}{dt} + V_R \tag{1.1}$$

Rearrange this equation such as

$$L\frac{di_R}{dt} = V_{CR} - V_R - Ri_R \tag{1.2}$$

On forging the same equations for S and T phases, we get

$$L\frac{di_S}{dt} = V_{CS} - V_S - Ri_S \tag{1.3}$$

$$L\frac{di_T}{dt} = V_{CT} - V_T - Ri_T \tag{1.4}$$

Putting Eqs. (1.2), (1.3), and (1.4) together, we have

$$\frac{d}{dt}\begin{bmatrix} i_R \\ i_S \\ i_T \end{bmatrix} = \frac{1}{L}\begin{bmatrix} V_{CR} \\ V_{CS} \\ V_{CT} \end{bmatrix} - \frac{1}{L}\begin{bmatrix} V_R \\ V_S \\ V_T \end{bmatrix} - \frac{R}{L}\begin{bmatrix} i_R \\ i_S \\ i_T \end{bmatrix} \tag{1.5}$$

The purpose of the dq-transformation is to transfer a parameter in ABC coordinates to a new coordinated called dq, where d stands for direct and q stands for quadrature [7, 8]. The advantage of this transformation technique is that, under a balance three-phase system, a three-phase ac parameter in ABC coordinates can be represented by a two-phase dc parameter in dq. These dc parameters are used for reducing the complication of mathematical calculations. After this, inverse transformation is performed for obtaining the actual three-phase ac quantities. Thus, dq-transformation method is used in any balanced three-phase stationary abc coordinate quantities (e.g., voltages, currents, etc.) by rotating with speed of fundamental angular frequency as shown below in the matrix form:

$$\begin{bmatrix} i_d \\ i_q \\ 0 \end{bmatrix} = K \begin{bmatrix} i_R \\ i_S \\ i_T \end{bmatrix} \text{ and } \begin{bmatrix} i_R \\ i_S \\ i_T \end{bmatrix} = K^{-1} \begin{bmatrix} i_d \\ i_q \\ 0 \end{bmatrix} \tag{1.6}$$

where

$$K^{-1} = \begin{bmatrix} \cos\omega t & \sin\omega t & 1 \\ \cos(\omega t - \frac{2\pi}{3}) & \sin(\omega t - \frac{2\pi}{3}) & 1 \\ \cos(\omega t + \frac{2\pi}{3}) & \sin(\omega t + \frac{2\pi}{3}) & 1 \end{bmatrix}$$

i_d is the active component of current and i_q is the reactive component of current. Here the zero-sequence components of line voltage and currents are neglected because of balanced three-phase source voltages [4, 6, 7]. Therefore two columns of K^{-1} matrix are considered for further calculation. Then,

$$T^{-1} = \begin{bmatrix} \cos\omega t & \sin\omega t \\ \cos(\omega t - \frac{2\pi}{3}) & \sin(\omega t - \frac{2\pi}{3}) \\ \cos(\omega t + \frac{2\pi}{3}) & \sin(\omega t + \frac{2\pi}{3}) \end{bmatrix} \tag{1.7}$$

For an inverse dq-transformation, Eq. (1.8) is used:

$$\begin{bmatrix} i_R \\ i_S \\ i_T \end{bmatrix} = T^{-1} \begin{bmatrix} i_d \\ i_q \end{bmatrix} \tag{1.8}$$

Similarly we can get the d and q components for the voltages as shown:

$$\begin{bmatrix} e_d \\ e_q \end{bmatrix} = T \begin{bmatrix} V_R \\ V_S \\ V_T \end{bmatrix} \tag{1.9}$$

and

$$\begin{bmatrix} V_d \\ V_q \end{bmatrix} = T \begin{bmatrix} V_{CR} \\ V_{CS} \\ V_{CT} \end{bmatrix} \tag{1.10}$$

Equation (1.8) is differentiated with time, then we get

$$\frac{d}{dt}\begin{bmatrix} i_R \\ i_S \\ i_T \end{bmatrix} = \frac{d}{dt}(T^{-1}\begin{bmatrix} i_d \\ i_q \end{bmatrix}) = \frac{dT^{-1}}{dt}\begin{bmatrix} i_d \\ i_q \end{bmatrix} + T^{-1}\begin{bmatrix} \frac{di_d}{dt} \\ \frac{di_q}{dt} \end{bmatrix} \tag{1.11}$$

Equations (1.9), (1.10), and (1.11) are used in Eq. (1.5). Then

$$\frac{dT^{-1}}{dt}\begin{bmatrix} i_d \\ i_q \end{bmatrix} + T^{-1}\begin{bmatrix} \frac{di_d}{dt} \\ \frac{di_q}{dt} \end{bmatrix} = \frac{1}{L}T^{-1}\begin{bmatrix} V_d \\ V_q \end{bmatrix} - \frac{1}{L}T^{-1}\begin{bmatrix} e_d \\ e_q \end{bmatrix} - \frac{R}{L}T^{-1}\begin{bmatrix} i_d \\ i_q \end{bmatrix} \tag{1.12}$$

This equation is multiplied by T. Then we get

$$T\frac{dT^{-1}}{dt}\begin{bmatrix} i_d \\ i_q \end{bmatrix} + TT^{-1}\begin{bmatrix} \frac{di_d}{dt} \\ \frac{di_q}{dt} \end{bmatrix} = \frac{1}{L}TT^{-1}\begin{bmatrix} V_d \\ V_q \end{bmatrix} - \frac{1}{L}TT^{-1}\begin{bmatrix} e_d \\ e_q \end{bmatrix} - \frac{R}{L}TT^{-1}\begin{bmatrix} i_d \\ i_q \end{bmatrix} \tag{1.13}$$

We know that [8]

$$TT^{-1} = I \quad \& \quad T\frac{dT^{-1}}{dt} = \begin{bmatrix} 0 & -\omega \\ \omega & 0 \end{bmatrix} \tag{1.14}$$

where I is an identity matrix.

Putting Eq. (1.14) in Eq. (1.13), then

$$\begin{bmatrix} \frac{di_d}{dt} \\ \frac{di_q}{dt} \end{bmatrix} = \begin{bmatrix} -\frac{R}{L} & \omega \\ -\omega & -\frac{R}{L} \end{bmatrix}\begin{bmatrix} i_d \\ i_q \end{bmatrix} + \frac{1}{L}\begin{bmatrix} V_d \\ V_q \end{bmatrix} - \frac{1}{L}\begin{bmatrix} e_d \\ e_q \end{bmatrix} \tag{1.15}$$

The differential equations for all phases R, S, and T in Fig. 1.1 can be found by practicing the KCL at the nodes X, Y, and Z:

$$C\frac{dV_{CR}}{dt} = i_{SR} - i_R \tag{1.16}$$

$$C\frac{dV_{CS}}{dt} = i_{SS} - i_S \tag{1.17}$$

$$C\frac{dV_{CT}}{dt} = i_{ST} - i_T \tag{1.18}$$

Equation (1.19) is obtained by putting the rearranged Eqs. (1.16), (1.17), and (1.18) into appropriate matrix form:

$$\frac{d}{dt}\begin{bmatrix} V_{CR} \\ V_{CS} \\ V_{CT} \end{bmatrix} = \frac{1}{C}\begin{bmatrix} i_{SR} \\ i_{SS} \\ i_{ST} \end{bmatrix} - \frac{1}{C}\begin{bmatrix} i_R \\ i_S \\ i_T \end{bmatrix} \tag{1.19}$$

After transforming abc-rotating frame into dq-stationary frame:

$$\begin{bmatrix} V_d \\ V_q \end{bmatrix} = T\begin{bmatrix} V_{CR} \\ V_{CS} \\ V_{CT} \end{bmatrix} \Rightarrow \frac{d}{dt}\begin{bmatrix} V_{CR} \\ V_{CS} \\ V_{CT} \end{bmatrix} = \frac{d}{dt}\left(T^{-1}\begin{bmatrix} V_d \\ V_q \end{bmatrix}\right)$$

$$= \frac{dT^{-1}}{dt}\begin{bmatrix} V_d \\ V_q \end{bmatrix} + T^{-1}\begin{bmatrix} \frac{dV_d}{dt} \\ \frac{dV_q}{dt} \end{bmatrix} \tag{1.20}$$

Putting Eq. (1.20) into Eq. (1.19), then we can get

$$\frac{dT^{-1}}{dt}\begin{bmatrix} V_d \\ V_q \end{bmatrix} + T^{-1}\begin{bmatrix} \frac{dV_d}{dt} \\ \frac{dV_q}{dt} \end{bmatrix} = \frac{1}{C}T^{-1}\begin{bmatrix} i_{Sd} \\ i_{Sq} \end{bmatrix} - \frac{1}{C}T^{-1}\begin{bmatrix} i_d \\ i_q \end{bmatrix} \tag{1.21}$$

Multiplying by T on both side Eq. (1.19)

$$T\frac{dT^{-1}}{dt}\begin{bmatrix} V_d \\ V_q \end{bmatrix} + TT^{-1}\begin{bmatrix} \frac{dV_d}{dt} \\ \frac{dV_q}{dt} \end{bmatrix} = \frac{1}{C}TT^{-1}\begin{bmatrix} i_{Sd} \\ i_{Sq} \end{bmatrix} - \frac{1}{C}TT^{-1}\begin{bmatrix} i_d \\ i_q \end{bmatrix} \tag{1.22}$$

Putting Eq. (1.14) in Eq. (1.22), we get

$$\begin{bmatrix} \frac{dV_d}{dt} \\ \frac{dV_q}{dt} \end{bmatrix} = \begin{bmatrix} 0 & \omega \\ -\omega & 0 \end{bmatrix}\begin{bmatrix} V_d \\ V_q \end{bmatrix} + \frac{1}{C}\begin{bmatrix} i_{Sd} \\ i_{Sq} \end{bmatrix} - \frac{1}{C}\begin{bmatrix} i_d \\ i_q \end{bmatrix} \tag{1.23}$$

Here CSC topology is controlled by the tri-level SPWM switching functions S_R, S_S, and S_T, where, for example,

$$S_R = \left\{ \begin{array}{lll} 1 & \rightarrow & \text{The switch } S_1 \text{ is } ON, \text{ and } S_2 \text{ is } OFF \\ -1 & \rightarrow & \text{The switch } S_2 \text{ is } ON, \text{ and } S_1 \text{ is } OFF \\ 0 & \rightarrow & S_1 \text{ and } S_2, \text{ both are } ON \text{ or both are } OFF \end{array} \right\}$$

and $S_R = S_S = S_T = 0$ means the shoot-through in one of the converter legs [9]. Therefore CSC controlling by the tri-level SPWM technique can be modeled as [1]

$$\begin{aligned} i_{SR} &= m_R I_{dc}; \\ i_{SS} &= m_S I_{dc}; \\ i_{ST} &= m_T I_{dc}. \end{aligned} \tag{1.24}$$

where m_R, m_T, and m_S symbolize the modulating signals of the three phases normalized to the peak of the triangular carrier signal. The modulating signal is usually a three-phase sinusoidal wave [10, 13, 14]. In literature [11, 15, 16], modulating signal is the ratio of the peak value of the elementary component to the magnitude of the pulsed waveform. The dq transformation is also allowed for the modulating signals. The d-q components of the transformed modulating signals are given below:

$$\begin{bmatrix} i_{Sd} \\ i_{Sq} \end{bmatrix} = \begin{bmatrix} M_d \\ M_q \end{bmatrix} I_{dc} \tag{1.25}$$

where M_d and M_q are the d- and q-axis components of the system modulating signals (m_R, m_T, and m_S), respectively [12, 17].

Using the power balance of the voltage and current quantities in the dc side and the ac side of the converter, we get

$$V_{dc} I_{dc} = -\frac{3}{2}\left(V_d i_{sd} + V_q i_{sq}\right) \tag{1.26}$$

$$\left(L_{dc}\frac{d}{dt}I_{dc} + R_{dc}I_{dc}\right) I_{dc} = -\frac{3}{2}\left(V_d i_{sd} + V_q i_{sq}\right) \tag{1.27}$$

Equation (1.27) is rearranged and then

$$\left(L_{dc}\frac{d}{dt}I_{dc} + R_{dc}I_{dc}\right) = -\frac{3}{2}\left(V_d \frac{i_{sd}}{I_{dc}} + V_q \frac{i_{sq}}{I_{dc}}\right) \tag{1.28}$$

Putting Eq. (1.25) in Eq. (1.28), then

$$L_{dc}\frac{d}{dt}I_{dc} + R_{dc}I_{dc} = -\frac{3}{2}\left(V_d M_d + V_q M_q\right) \tag{1.29}$$

Equation (1.25) is used in Eq. (1.23). Then

$$\begin{bmatrix} \frac{dV_d}{dt} \\ \frac{dV_q}{dt} \end{bmatrix} = \begin{bmatrix} 0 & \omega \\ -\omega & 0 \end{bmatrix} \begin{bmatrix} V_d \\ V_q \end{bmatrix} + \frac{1}{C}\begin{bmatrix} M_d \\ M_q \end{bmatrix} I_{dc} - \frac{1}{C}\begin{bmatrix} i_d \\ i_q \end{bmatrix} \tag{1.30}$$

Equations (1.15), (1.29), and (1.30) are rearranged, and then the CSC-STATCOM model in matrix form is obtained:

$$\frac{d}{dt}\begin{bmatrix} I_{dc} \\ i_d \\ i_q \\ V_d \\ V_q \end{bmatrix} = \begin{bmatrix} -\frac{R_{dc}}{L_{dc}} & 0 & 0 & -\frac{3}{2}\frac{M_d}{L_{dc}} & -\frac{3}{2}\frac{M_q}{L_{dc}} \\ 0 & -\frac{R}{L} & \omega & \frac{1}{L} & 0 \\ 0 & -\omega & -\frac{R}{L} & 0 & \frac{1}{L} \\ \frac{1}{C}M_d & -\frac{1}{C} & 0 & 0 & \omega \\ \frac{1}{C}M_q & 0 & -\frac{1}{C} & -\omega & 0 \end{bmatrix} * \begin{bmatrix} I_{dc} \\ i_d \\ i_q \\ V_d \\ V_q \end{bmatrix} + \begin{bmatrix} 0 \\ -\frac{1}{L}e_d \\ -\frac{1}{L}e_q \\ 0 \\ 0 \end{bmatrix} \tag{1.31}$$

In Eq. (1.31) M_d and M_q denote the input variables. I_{dc} and i_q denote the output variables. Here, ω denotes the frequency at which the system rotates and is equivalent to the system voltages bottom level frequency. i_d and i_q are the d-axis and q-axis components of the line current. V_d and V_q are the d-axis and q-axis components of the voltage across filter capacitor, respectively. From Eq. (1.31), it can be observed that the reason behind the nonlinearity of the CSC-STATCOM design is the I_{dc} segment. So this annoying nonlinearity can be eliminated by accurate CSC based STATCOM modeling. Nonlinearity of this modeling is eliminated with the assistance of power balance equation. At this point, it is presumed that the power loss in the resistance R_{dc} and switches can be neglected and the shunt transformers turns ratio is assumed to be $n : 1$.

As shown in Fig. 1.2, the new coordinate system is defined where the d-axis always concurrent with the vector representing instantaneous voltage. By outlining the d-axis to be always concurrent with the instantaneous voltage vector E, this transformation yields E_d equals E, while E_q is set at zero, i.e., line voltages are written in dq form such as $[E] = [E_d 0]^T$ [18, 19]. From power system principles, we assumed that [1, 20, 21]

$$\begin{bmatrix} e_d \\ e_q \end{bmatrix} = \begin{bmatrix} \frac{E_d}{n} \\ 0 \end{bmatrix} \tag{1.32}$$

The active power delivery by the ac source and the dc-side active power absorption can be stated as

$$P_{ac} = -\frac{3}{2} * \frac{E_d}{n} i_d;\ P_{dc} = V_{dc} I_{dc} \Rightarrow P_{dc} = L_{dc} I_{dc} \frac{d}{dt} I_{dc} + R_{dc} {I_{dc}}^2$$

The relationship between P_{ac} and P_{dc} is $P_{ac} = P_{dc} + P_{\text{loss}}$, where P_{loss} is the power loss in the resistor R. The resistance R is always very small; hence, it is practically reasonable to neglect its power loss without noticeable loss of accuracy. From $P_{ac} = P_{dc}$, the following dynamic equation results:

$$-\frac{3}{2} * \frac{E_d}{n} i_d = L_{dc} I_{dc} \frac{d}{dt} I_{dc} + R_{dc} {I_{dc}}^2 \tag{1.33}$$

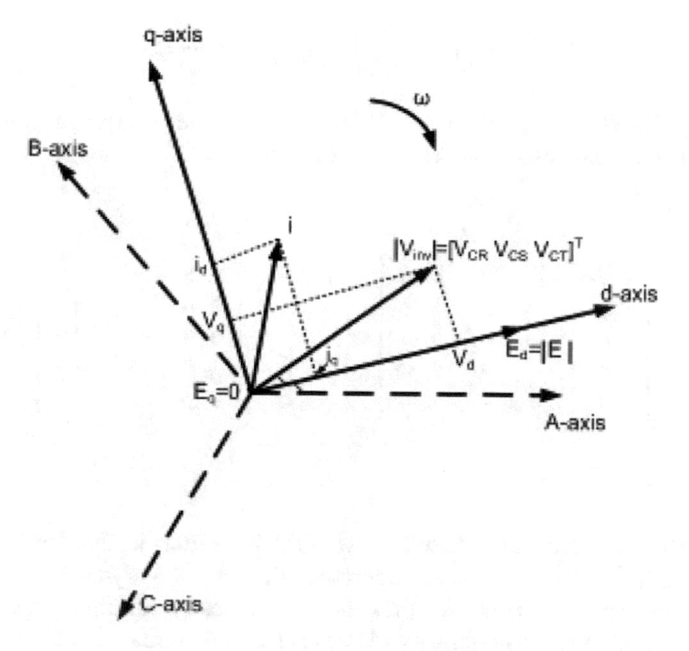

Fig. 1.2 The synchronous rotating d-q reference frame description

which can be rewritten as

$$I_{dc}\frac{d}{dt}\left(I_{dc}\right) = -\frac{R_{dc}}{L_{dc}}\left(I_{dc}^{2}\right) - \frac{3}{2}\frac{E_d}{nL_{dc}}i_d \tag{1.34}$$

After using power balance equation and mathematical manipulation, nonlinearity is removed. Finally the Eq. (1.34) is obtained which can be rewritten as:

$$\frac{d}{dt}\left(I_{dc}^{2}\right) = -\frac{2R_{dc}}{L_{dc}}\left(I_{dc}^{2}\right) - \frac{3E_d}{L_{dc}n}i_d \tag{1.35}$$

In Eq. (1.35) state variable (I_{dc}) is replaced by the state variable (I_{dc}^2), to make the dynamic equation linear. Thus, the improved better dynamic model of the CSC-STATCOM is obtained and given in the Eqs. (1.36), (1.37), (1.38), (1.39), and (1.40).

$$\frac{di_d}{dt} = -\frac{R}{L}i_d + \omega i_q + \frac{1}{L}V_d - \frac{1}{L}e_d \tag{1.36}$$

$$\frac{di_q}{dt} = -\omega i_d - \frac{R}{L}i_q + \frac{1}{L}V_q \tag{1.37}$$

$$\frac{dV_d}{dt} = \omega V_q + \frac{1}{c}i_{Sd} - \frac{1}{c}i_d \tag{1.38}$$

$$\frac{dV_q}{dt} = -\omega V_d + \frac{1}{c}i_{Sq} - \frac{1}{c}i_q \tag{1.39}$$

$$\frac{d}{dt}\left(I_{dc}^2\right) = -\frac{2R_{dc}}{L_{dc}}\left(I_{dc}^2\right) - \frac{3E_d}{L_{dc}n}i_d \tag{1.40}$$

Finally Eqs. (1.36), (1.37), (1.38), (1.39), and (1.40) are rearranged, and then the better dynamic and robust model of the CSC-SATACOM in matrix form can be derived as

$$\frac{d}{dt}\begin{bmatrix} I_{dc}^2 \\ i_d \\ i_q \\ V_d \\ V_q \end{bmatrix} = \begin{bmatrix} -\frac{2R_{dc}}{L_{dc}} & -\frac{3E_d}{nL_{dc}} & 0 & 0 & 0 \\ 0 & -\frac{R}{L} & \omega_o & \frac{1}{L} & 0 \\ 0 & -\omega & -\frac{R}{L} & 0 & \frac{1}{L} \\ 0 & -\frac{1}{C} & 0 & 0 & \omega \\ 0 & 0 & -\frac{1}{C} & -\omega & 0 \end{bmatrix} * \begin{bmatrix} I_{dc}^2 \\ i_d \\ i_q \\ V_d \\ V_q \end{bmatrix} + \begin{bmatrix} 0 & 0 \\ 0 & 0 \\ 0 & 0 \\ \frac{1}{C} & 0 \\ 0 & \frac{1}{C} \end{bmatrix} * \begin{bmatrix} i_{Sd} \\ i_{Sq} \end{bmatrix} + \begin{bmatrix} 0 \\ -\frac{1}{nL} \\ 0 \\ 0 \\ 0 \end{bmatrix} * E_d \tag{1.41}$$

Above modeling of CSC based STATCOM is written in the form of modern control methods, i.e., state-space representation. In the dynamic modeling of systems, state-space equation requires the three types of variables: state variables (x), input variables (u), and output variables (y) with disturbance (e). So comparing Eq. (1.41) with the standard state-space representation, i.e.,

$$\dot{x} = Ax + Bu + Fe \tag{1.42}$$

$$y = Cx \tag{1.43}$$

where
A = 5 × 5 constant matrix; B = 5 × 2 constant matrix; C = 2 × 5 constant matrix; F = 5 × 1 constant matrix;

$$x = \left[I_{dc}^2\ i_d\ i_q\ V_d\ V_q\right]^T \tag{1.44}$$

$$u = \left[i_{Sd}\ i_{Sq}\right]^T \tag{1.45}$$

$$y = \left[I_{dc}^2\ i_q\right]^T \tag{1.46}$$

$$e = E_d$$

$$A = \begin{bmatrix} -\frac{2R_{dc}}{L_{dc}} & -\frac{3E_d}{nL_{dc}} & 0 & 0 & 0 \\ 0 & -\frac{R}{L} & \omega & \frac{1}{L} & 0 \\ 0 & -\omega & -\frac{R}{L} & 0 & \frac{1}{L} \\ 0 & -\frac{1}{c} & 0 & 0 & \omega \\ 0 & 0 & -\frac{1}{c} & -\omega & 0 \end{bmatrix}; B = \begin{bmatrix} 0 & 0 \\ 0 & 0 \\ 0 & 0 \\ \frac{1}{c} & 0 \\ 0 & \frac{1}{c} \end{bmatrix}; C = \begin{bmatrix} 1 & 0 \\ 0 & 0 \\ 0 & 1 \\ 0 & 0 \\ 0 & 0 \end{bmatrix}^T; F = \begin{bmatrix} 0 \\ -\frac{1}{nL} \\ 0 \\ 0 \\ 0 \end{bmatrix}$$

In Eqs. (1.42) and (1.43), two control inputs (i_{sd} and i_{sq}), two control outputs (I_{dc}^2 and iq), and five system states (I_{dc}^2, i_d, i_q, V_d, and V_q) are presented. Circuit parameters which are used in Eq. (1.41) are selected as below:

$$R = 0.01\Omega;\ L = 1.86 \times 10^{-3}\text{H};\ C = 350 \times 10^{-6}\text{F};\ \omega = 314;\ R_{dc} = 0.25\Omega;$$
$$L_{dc} = 37.5 \times 10^{-3}\text{H}$$

After this, in order to check the stability of the system, the eigenvalues of the linearized test system are found. Since all the systems eigenvalues exist on the left side of the system plane, i.e., the system is stable. After this, for checking the observability and controllability of the system, the state-space model of CSC-STATCOM has five state variables. All of them can be measured from the power system, which means the system is observable. In our case, the state matrix $A \in R^{5*5}$ and the input matrix $B \in R^{5*5}$, we need to find the rank of matrix C_o for investigation of system controllability. If the matrix C_o has full rank, then system will be controllable. Controllability matrix (C_o) can be written as

$$C_o = \left[A : AB : AAB\right] \tag{1.47}$$

The values of A and B are placed in Eq. (1.47). The rank of controllability matrix (C_o) is found as follows:

$$\det(C_o) = 1.3521\text{e} + 024 \qquad \text{and} \qquad \text{rank}\,(\text{ctrb(A, B)}) = 5$$

The matrix C_o has full rank, i.e., 5; thus the linearized CSC-STATCOM model is controllable. Now, find the open-loop responses of input and output variables of the system. These are shown in Figs. 1.3, 1.4, and 1.5 with many big oscillations in iq and id when 0–1.5 s i_q, i_d, and I_{dc}^2 is not stable.

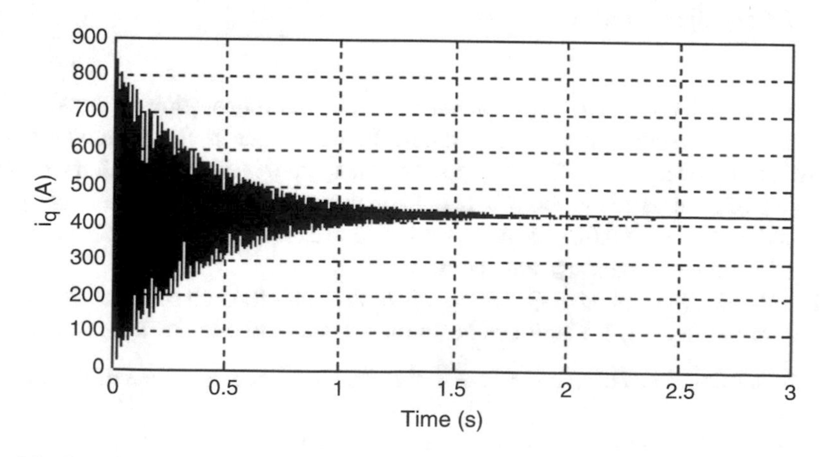

Fig. 1.3 Open-loop i_q outcome

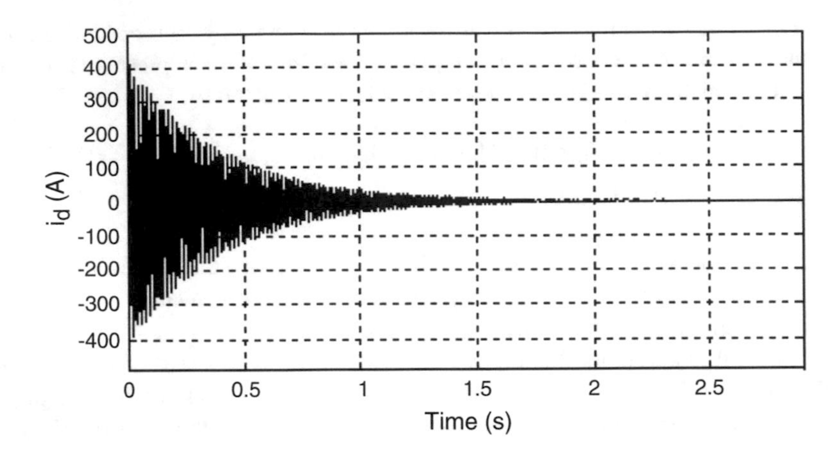

Fig. 1.4 Open-loop i_d outcome

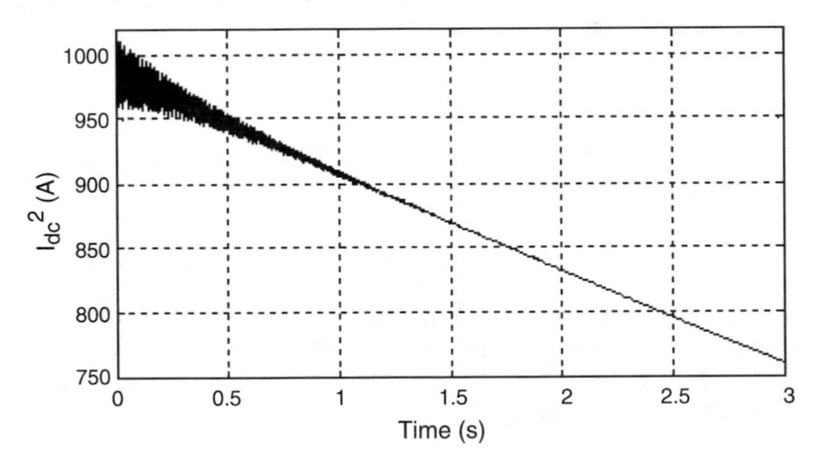

Fig. 1.5 Open-loop I_{dc}^2 outcome

Since the STATCOM is used for reactive power compensation in transmission systems, the transient response of i_q should be within one cycle of time period of transmission systems [22]. The frequency of the system is 50 Hz, which means that one cycle of time period is about 0.02 s. Thus the transient response time of i_q and i_d should be less than 0.02 s., and the dc-side current (I_{dc}) should remain constant during this time for stable system operation.

The characteristic plots of i_q, i_d, and I_{dc}^2 of the open-loop response of the system are shown in Figs. 1.3, 1.4, and 1.5. From these characteristics we observe that:

(a) The settling times of i_d and i_q are higher than the necessary time of 0.02 s.
(b) The initial overshoots of i_d and i_q are very high.
(c) The dc-side current (I_{dc}^2) is monotonically decreasing.

Hence it is observed that the open-loop response of the system, as shown in Figs. 1.3, 1.4, and 1.5, is undesirable, and these dynamic characteristics of the system must be improved in order to achieve the better stability of the system.

Above problems are removed by using the different controllers. Presently the most used techniques for controller design of FACTS devices are the proportional integration (PI), PID controller [23], pole-shifting controller, and linear quadratic regulator (LQR) [22]. But LQR and pole-shifting algorithms give fast response in comparison to PI and PID algorithm [24]. The pole-shifting and LQR controllers are based on dynamic modeling of CSC based STATCOM. Pole-shifting method is one of the classic control theories. Theoretically pole-shifting technique is to set the desired pole location of system and to move the current pole location of the system to that desired pole location in order to get the desired system response. LQR is also able to give a fast response. But in LQR algorithm, the desired pole location is obtained with the help of two cost functions. Now, the design of dc-link reactor for the CSC-STATCOM will be discussed, before checking the compatibility of different controllers for CSC-STATCOM.

1.3 Design of DC-Link Reactor

DC-link reactor is chosen in view of the CSC transient response, total demand distortion (TDD) of the CSC line current waveforms, power loss, and reactor cost. In the transient state, since the reactive power produced by CSC is proportional to the mean dc-link current, transition from one operating point to another needs large amount of energy to be injected or extracted from the dc link, i.e.,

$$\text{Energy} = (1/2)\, L_{dc}\left[I^2_{dc(\text{final})} - I^2_{dc(\text{initial})}\right] \tag{1.48}$$

From this viewpoint, a small L_{dc} is to be chosen. The value of dc-link inductance affects the response time of CSC-STATCOM, and the turn-off current of the power semiconductor devices leads for the system losses. Since the dc-link reactor is subject to short-duration voltage pulses, during each pulse, the dc-link reactor is considered to be operating in subtransient state. Therefore, the voltage-current equation of dc-link reactor given in Eq. (1.49) can be approximated to the first-order linear differential equation in Eq. (1.50):

$$\left(R_{dc}i_{dc}(t) + L_{dc}\frac{d}{dt}i_{dc}(t)\right) = V_{dc} \tag{1.49}$$

$$L_{dc}\frac{d}{dt}i_{dc}(t) \cong V_{dc} \tag{1.50}$$

where V_{dc} is the applied step voltage. The solution of the differential Eq. (1.50) is

$$i_{dc}(t) = \left(\frac{V_{dc}}{L_{dc}}\right)t + I_o \tag{1.51}$$

where I_0 is the initial value of the dc-link current at $t = 0$, by assuming that a step voltage is applied at that instant. Therefore, the dc-link current rises and decays approximately on a straight line in the steady state. $\frac{V_{dc}}{L_{dc}}$ dictates the rate of rise of the current. It is much smaller than the maximum permissible value of semiconductor switches di/dt rating (maximum value:1000 A/s). In view of this design constraint, a much smaller value of dc-link inductance could be used in the design.

The solution of Eq. (1.49) is given in the following:

$$i_{dc}(t) = \left(\frac{V_{dc}}{L_{dc}}\right)\left(1 - e^{-t/\tau_{dc}}\right) \tag{1.52}$$

where $\tau_{dc} = \frac{L_{dc}}{R_{dc}}$.

For the system, $\tau_{dc} = 0.15$, and hence theoretically in 5dc time period, dc-link current rises to $\frac{V_{dc}}{L_{dc}} = 75$ kA, under loss of control. During the charging period of the dc-link reactor, current rises to a rated value, theoretically, in a very short time period ($t_s = 2$ ms) [25]. However practically, the fully charging or discharging of dc-link reactor is limited by other circuit elements.

Internal resistance R_{dc} of the dc-link reactor should be kept at an implementable minimum value, thus reducing the dc reactor losses. Operating the dc-link reactor at a higher voltage for the same system size could make a more significant contribution to reactor loss reduction problem than R_{dc} reduction. Superconducting reactor could only make a contribution to loss reduction problem at the expense of much higher reactor costs.

From the viewpoint of distortion of line current waveforms at the input of CSC, large dc-link inductors are more desirable. This choice yields to better approximation of dc-link current to a level current, thus improving the TDD of the input current waveform. A large dc-link reactor means a huge and expensive system component; therefore, it is not recommended. A compromise is therefore needed for the designing of L_{dc} and R_{dc}.

In view of above considerations and specification of CSC based STATCOM given so far, inductance, L_{dc}, is found to be 37.5 mH, and R_{dc} = 0.25 is found to be adequate. In an air-core reactor, the only power loss is due to the copper losses. Since air-core reactor has the number of turns higher than that of an iron-core reactor due to its lower permeability, this results in higher losses than that of iron-core reactor. Although from the viewpoint of dc-link reactor losses, iron-core reactor is better than air-core reactor, dc-link reactor has been chosen as air-core reactor, which can be used at outdoor with natural air cooling.

Now in the following sections, the compatibility between different controllers and CSC based STATCOM for desired output of the system is checked.

1.4 Pole-Shifting Controller Design

The pole-shift technique is one of the basic control methods employed in feedback control system theory. Theoretically pole-shift technique is to set the preferred pole position and to move the pole position of the system to that preferred pole position, to get the desired system outcomes [26]. Here poles of system are shifted because the position of the poles related directly to the eigenvalues of the system, which control the dynamic characteristics of the system outcomes. But for this method, the system must be controllable. The pole-shifting controller design for CSC-STATCOM is explained here. For the state-space representation of the test system, Eqs. (1.42) and (1.43) are rewritten again as below:

$$x = Ax + Bu + Fe; \qquad y = Cx$$

where x is the state vector, u is the input vector, A is the basis matrix, B is the input matrix, and e is the disturbance input.

If the controller is set as

$$u = -Kx + Ty_{\text{ref}} + Me \tag{1.53}$$

where $$y_{\text{ref}} = \left[I^2_{dc(\text{ref})} \ I_{q(\text{ref})} \right]^T$$

then the state equation of closed loop can be written as

$$\dot{x} = (A - BK)x + BTy_{\text{ref}} + BMe + Fe \tag{1.54}$$

where $T = (C*(-(A-B*K)^{-1})*B)^{-1}$ and $M = (C*(-A+B*K)^{-1})^{-1}*(-C*(-A + B*K)^{-1} * F)$; these values are found out from mathematical calculation [27, 28]. Here, state-feedback gain matrix is represented by K. The gain matrix K is designed in such a way that Eq. (1.55) is satisfied with the desired poles:

$$\left| sI - (A - BK) \right| = (s - P_1)(s - P_2) \ldots\ldots\ldots (s - P_n) \tag{1.55}$$

where $P1, P2, \ldots Pn$ are the desired pole locations. Equation (1.55) is the desired characteristic polynomial equation. The values of $P1, P2, \ldots Pn$ are selected such that the system becomes stable and all closed-loop eigenvalues are located in the left half of the complex plane. The final configuration of the pole-shifting controller-based CSC-STATCOM is shown in Fig. 1.6.

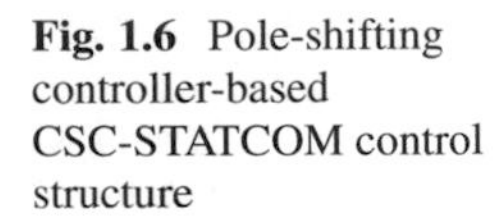

Fig. 1.6 Pole-shifting controller-based CSC-STATCOM control structure

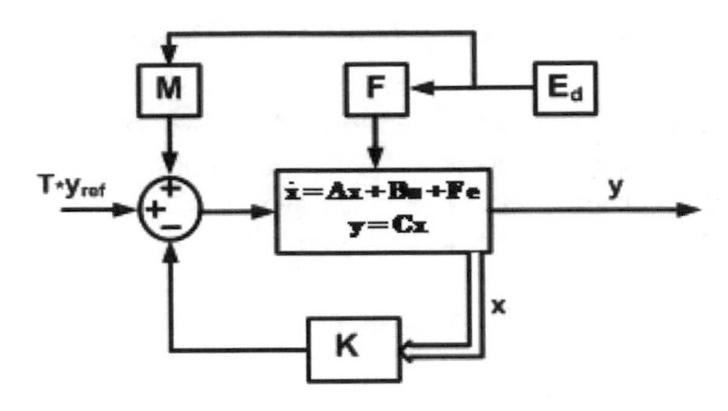

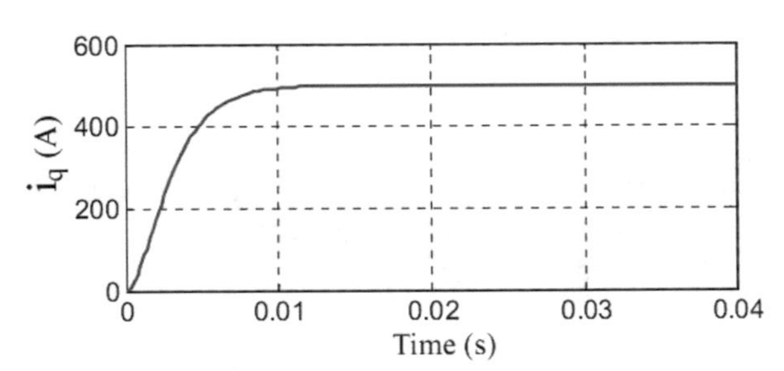

Fig. 1.7 i_q response with pole-shifting method

1.4.1 Simulation Results Using Pole-Shifting Method

In this study, the model of CSC-STATCOM and controllers are developed in MATLAB/Simulink environment. Circuit parameters which are used in Fig. 1.1 are selected as below:

$$R = 0.01\Omega;\ L = 1.86 \times 10^{-3}\text{H};\ C = 350 \times 10^{-6}\text{F};\ \omega = 314;\ R_{dc} = 0.25\Omega;$$

$$L_{dc} = 37.5 \times 10^{-3}\text{H}$$

According to problems in open-loop responses, we want to decrease the settling time of all three states (i_q, i_d, and I_{dc}^2) and reduce the steady-state error of I_{dc}^2 further such that the requirements are satisfied. After trial and error method, we set the poles on [−650 −650 −600 −600 −540]. The system response is shown in Figs. 1.7, 1.8, and 1.9. The settling time for all three states is decreased to less than 0.02 s and I_{dc}^2 is 1000 A.

Pole locations in the pole-shifting controller must be chosen with carefully and wisely, which needs broad knowledge of the system characteristics. Due to this situation, the value of feedback gain matrix (K) is not a unique solution by the resulting equations. Thus, MIMO system designing by pole-shifting method is a difficult task. In MIMO system designing, feedback gain must be high for reducing the system response time and eliminating the disturbances from the system. But high feedback gain value gives rise to high input that is restricted by the hardware. Thus for MIMO systems, we need to make a compromise between the input energy and

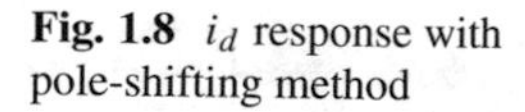
Fig. 1.8 i_d response with pole-shifting method

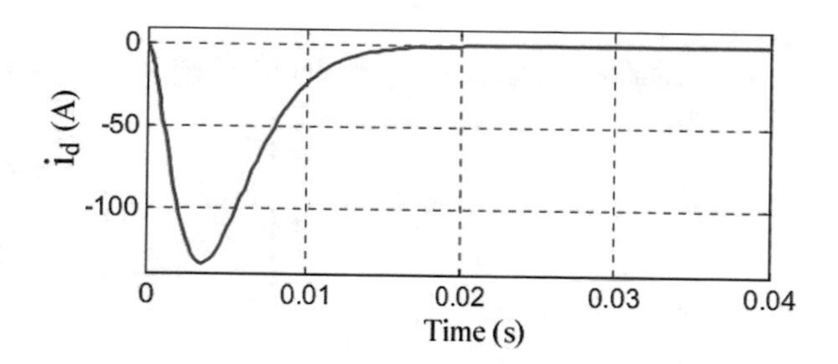

Fig. 1.9 I_{dc}^2 response with pole-shifting method

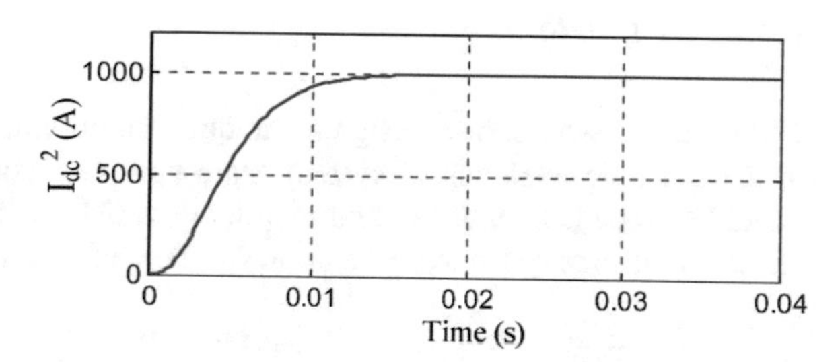

output performance and pay lots of attention while selecting poles. It is still tough to search the optimal pole location which can give best possible solutions according to our requirements.

By using LQR method, we can utilize all the liberty of MIMO system and achieve good cooperation between the utilization of system performance and control effort and can also promise a stable and efficient system. In LQR method, the specific values of Q and R matrices are needed for the minimization of cost function (J_C). But it is comparatively easy to attain all the requirements. A unique solution of state-feedback matrix K is obtained by the minimum value of cost function J_C. Therefore in the following sections, LQR controller with CSC-STATCOM is explained and analyzed.

1.5 Linear Quadratic Regulator for CSC-STATCOM

Another MIMO design method in the optimal control techniques is linear quadratic regulator (LQR), which is operated a dynamic system at minimum cost. LQR theory is no doubt one of the most basic control system design methods. LQR algorithm is the optimal theory of pole placement method and picks the best possible pole location based on the cost function, J_C. This method finds the optimal gain matrix that minimizes J_C in order to achieve some compromise between the use of control effort, the magnitude, and the speed of response that will give the guarantee of a stable system.

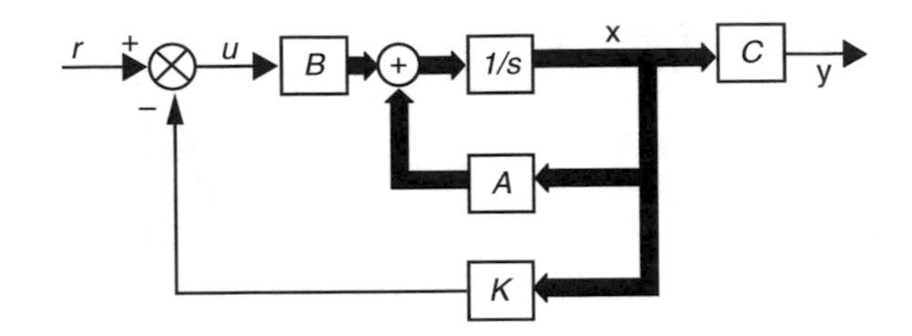

Fig. 1.10 Classic representation of state space regulator

1.5.1 Algorithm

The LQR is a well-known design technique that provides practical feedback gains. In the dynamic modeling of system, state-space equations involve three types of variables: state (x), input (u), and output (y) variables. The plant can be written in the classical state-space representation for the derivation of LQR method.

$$\dot{x} = Ax + Bu \tag{1.56}$$

$$y = Cx \tag{1.57}$$

with $x \in R^n$ the state variables relative to the system, $y \in R^m$ the output of the system, $u \in R^r$ the input of the system, $A \in R^{n*n}$ the state matrix, $B \in R^{n*r}$ the control input matrix, and $C \in R^{m*n}$ the output matrix. The system is regulated with a state-feedback controller (Fig. 1.10) with $r = 0$. Therefore

$$u(t) = -Kx(t) \tag{1.58}$$

with K the coefficients of the state feedback [29].

So in order to minimize the performance index, one way of representing the quadratic cost function in mathematics for LQR formulation is shown below:

$$J_c = \int_0^\infty x^T Qxdt + \int_0^\infty u^T Rudt \tag{1.59}$$

where Q is a positive-definite (or positive-semidefinite) Hermitian or real symmetric matrix and R is also a positive-definite Hermitian or real symmetric matrix. Note that the second term on the right-hand side of the Eq. (1.59) accounts for the expenditure of the energy of the control signals. The matrices Q and R determine the relative importance of the error and the expenditure of this energy.

$$J_c = \int_0^\infty x^T Qxdt + \int_0^\infty x^T K^T RKxdt = \int_0^\infty x^T \left(Q + K^T RK\right)xdt \tag{1.60}$$

From the above equations, we get

$$x^T \left(Q + K^T RK\right) x = -\frac{d}{dt} \left(x^T Px\right) \tag{1.61}$$

then we get

$$x^T \left(Q + K^T RK\right) x = -\dot{x}^T Px - x^T P\dot{x} = -x^T \left[(A - BK)^T P + P(A - BK)\right] x \tag{1.62}$$

By correlating both sides of the above equation, for the sake of satisfying the equation for any value of x, we require that

$$(A - BK)^T P + P(A - BK) = -(Q + K^T RK). \tag{1.63}$$

In Eq. (1.63), R is a real symmetric matrix or positive-definite Hermitian; we can write

$$R = T^T T \tag{1.64}$$

where T represents a nonsingular matrix and

$$A^T P + PA + \left[TK - (T^T)^{-1} B^T P\right]^T \left[TK - (T^T)^{-1} B^T P\right] - PBR^{-1} B^T P + Q = 0 \tag{1.65}$$

For obtaining the minimized cost function (J_c), Eq. (1.66) should be minimized.

$$x^T \left[TK - (T^T)^{-1} B^T P\right]^T \left[TK - (T^T)^{-1} B^T P\right] x. \tag{1.66}$$

Here this equation is nonnegative. The result is minimum when equation is zero or when

$$TK = (T^T)^{-1} B^T P. \tag{1.67}$$

Hence

$$K = T^{-1}(T^T)^{-1} B^T P = R^{-1} B^T P. \tag{1.68}$$

Thus control law is obtained as

$$u(t) = -Kx(t) = -R^{-1} B^T Px(t) \tag{1.69}$$

For solving the Eq. (1.69), the value of P must be known. So P is determined from the reduced Riccati equation:

$$A^T P + PA - PBR^{-1} B^T P + Q = 0 \ . \tag{1.70}$$

1.5.2 Controller Design

LQR controller can be easily designed to use above basic algorithm. LQR controller design for CSC based STATCOM; disturbance (e) is also added in the classical state-space representation. State-space modeling of CSC-STATCOM is already discussed in above section. Therefore in this part, the designing of LQR controller for CSC-STATCOM is explained and formulated. Now in the presence of disturbance, the control equation for LQR controller can be formulated as

$$u = -K * x + J * y_{\text{ref}} + N * e \tag{1.71}$$

Then the closed-loop state Eq. (1.42) is altered as

$$\dot{x} = (A - B * K) * x + B * J * y_{\text{ref}} + B * N * e + F * e \tag{1.72}$$

where K is the state-feedback matrix, N is the feedback matrix for the disturbance input, y_{ref} is the reference input, and J is a diagonal matrix which is used to obtain the unity steady-state gain. The matrix N should be designed in such a way that the output y should be minimally influenced by the disturbance (e). This objective can be achieved by making

$$\lim_{s \to 0} C(s * I - A + B * K)^{-1}(B * N + F) = 0 \tag{1.73}$$

$$\dot{x} = 0 \tag{1.74}$$

Here Eq. (1.73) represents the under steady-state condition. So, in steady-state condition, Eq. (1.43) can be expressed as

$$y = C(BK - A)^{-1}[BJ y_{\text{ref}} + (BN + F)e] \tag{1.75}$$

Therefore J and N values are found out from a mathematical calculation for tracking the reference output value (y_{ref}) by the system output value (y). For this, two conditions are used which are found from Eq. (1.75). These conditions are such as

$$\begin{cases} C(BK - A)^{-1}BJ = I \ , \\ C(BK - A)^{-1}(BN + F)e = 0 \end{cases} \tag{1.76}$$

where I is a 2*2 identify matrix. The values of J and N can be found using Eq. (1.76). These entries are such as

$$J = \left(C\left(-(A - BK)^{-1}\right)B\right)^{-1} \tag{1.77}$$

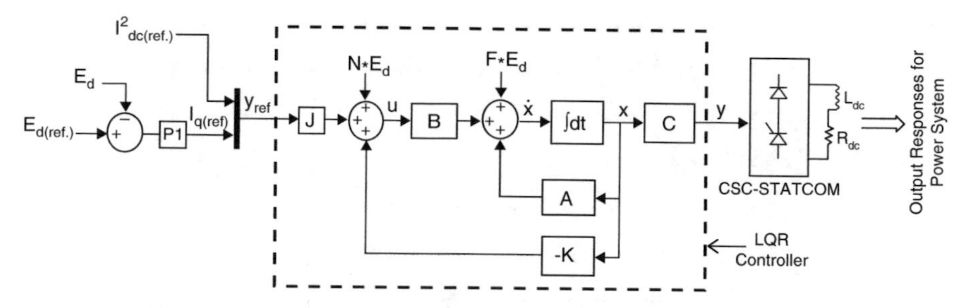

Fig. 1.11 Control structure of LQR based CSC-STATCOM

$$N = \left(\left(C(-A + BK)^{-1}B\right)\right)^{-1} * \left(-C(-A + BK)^{-1}F\right) \tag{1.78}$$

Here y_{ref} is given by

$$y_{\text{ref}} = \begin{bmatrix} I^2_{dc(\text{ref})} \\ I_{q(\text{ref})} \end{bmatrix} \tag{1.79}$$

In order to reduce the cost function (performance index) and to achieve the best value of gain matrices K, the following two equations are used which are derived in above section:

$$K = R^{-1}B^T P \qquad \text{and} \qquad A^T P + PA - PBR^{-1}B^T P + Q = 0$$

The final configuration of the LQR controller-based CSC-STATCOM is shown in Fig. 1.11. In this figure, $I_{q(\text{ref})}$ can be obtained from the PI controller output signal, and the input signal of the PI controller is the difference between system reference voltage ($E_{d(\text{ref})}$) and system voltage (E_d).

1.5.3 Simulation Results Using LQR Method

In this inspection, MATLAB/Simulink environment is employed to develop CSC based STATCOM and LQR controller model. The system circuit configuration is used as shown in Fig. 1.1, in pursuance of getting the better system response with reference to the minor overshoot, decreasing the settling time, and reducing the steady-state error, LQR controller is used for CSC-STATCOM. In this method, diagonal elements diag [Q] and diag [R] of matrices Q and R are initially selected. Many distinct sets of Q and R matrices for LQR controller are employed in the

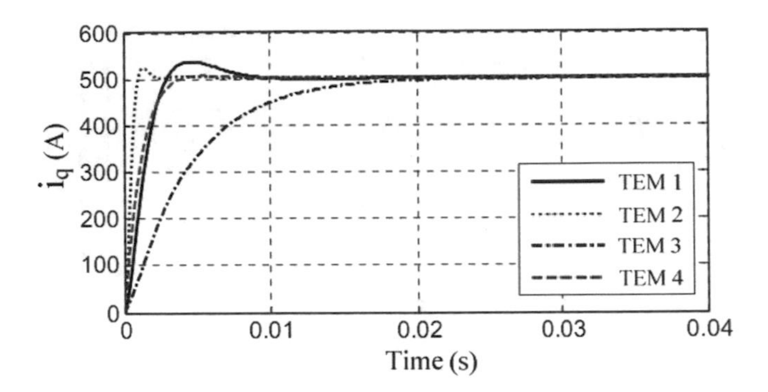

Fig. 1.12 Response of i_q for distinct LQR controllers (TEM 1, TEM 2, TEM 3, TEM 4)

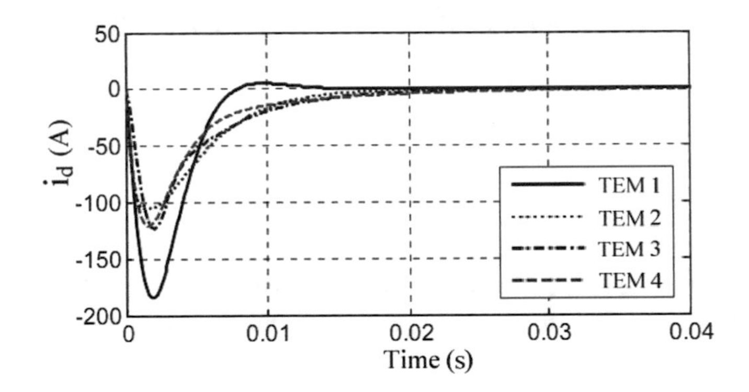

Fig. 1.13 Response of i_d for distinct LQR controllers (TEM 1, TEM 2, TEM 3, TEM 4)

system to check the steadiness of the system outcome with the assistance of trial and error method (TEM). For all three states (i_q, i_d, and I_{dc}^2), the system responses at distinct series of Q and R matrices are displayed in Figs. 1.12, 1.13, and 1.14. The distinct series of matrices Q and R for distinct LQR controllers are illustrated as:

Case 1 (TEM 1):-
diag [Q]:-[0.853, 18.593, 45.8, 13.2, 0.091, 10.3, 6.22] and diag [R]:-[0.341, 0.601]
Case 2 (TEM 2):-
diag [Q]:-[1, 9, 260, 55.1, 22.9, 23.1, 9.59] and diag [R]:- [6.9, 11.4]
Case 3 (TEM 3):-
diag [Q]:-[0.25, 9.9, 1.354, 0.075, 9.664, 2.55, 2.256] and diag [R]:-[3.14, 0.239]
Case 4 (TEM 4):-
diag [Q]:-[0.20, 21.55, 68.99, 23.33, 19.96, 21.095, 0.13] and diag [R]:-[0.394, 0.496]

From Figs. 1.12, 1.13, and 1.14, we can find that the response of i_q is perfect, but the I_{dc}^2 and i_d do not converge within 0.02 s in case 4. Case 3 represents the improved

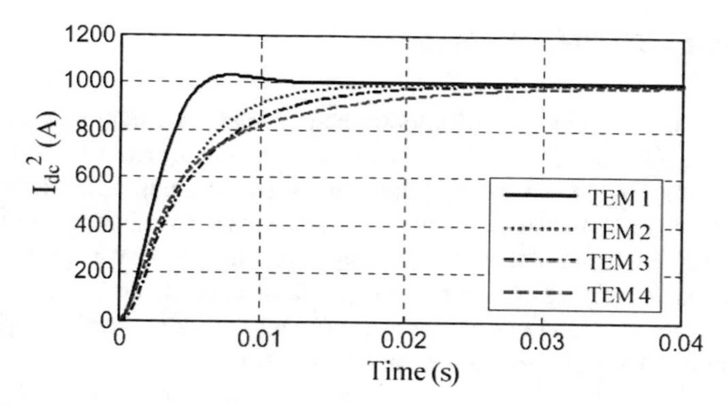

Fig. 1.14 Response of I_{dc}^2 for distinct LQR controllers (TEM 1, TEM 2, TEM 3, TEM 4)

results, but these results are not acceptable according to our requirements because overshoot and settling time of I_{dc}^2 and i_q are very high. In case 1, the response of I_{dc}^2 and i_q converges to the desired constant value with the required time, but the overshoot is high as shown in Table 1.2. From Figs. 1.12, 1.13, and 1.14, it can be seen that the settling time of I_{dc}^2, i_q, and i_d are within 0.015 s, which means they satisfy the design requirements in case 2. And the steady-state errors of I_{dc}^2 and i_q also satisfy the requirements. Among these cases, it can be said that case 2 (TEM2) is given best outcomes from Figs. 1.12, 1.13, and 1.14 and Table 1.2.

After seeing above results, we can say that the weight matrices of LQR controller have a main dominant role for designing of LQR based CSC-STATCOM. But these values are finding by the trial and error method which is time-consuming and laborious. Therefore, artificial intelligence (AI) techniques such as online PSO are employed for tuning of Q and R matrices in a very short time. AI techniques have been used to optimally find out the Q and R weight matrices for obtaining the optimized K matrix of the state-feedback gain with minimizing the cost function-based performance index. Now the controller is optimally designed through AI optimization techniques in the next section.

1.6 AI Technique-Based Proposed Controller Design

Basic definition of artificial intelligence (AI) is that it is a powerful method, particularly intelligent agent-based computer programs, by which machines work with intelligence using the latest engineering and science technologies. It is related to the similar task of using computers to understand human intelligence, but AI does not have to confine itself to methods that are biologically observable. AI technique, such as particle swarm optimization (PSO), has been applied successfully to design heuristic search algorithms [30]. The PSO-based AI technique is applied in the design of LQR controller, and this is explained as below.

1.6.1 PSO-Based LQR Controller

PSO is the nature-inspired heuristic optimization method which first proposed by Kennedy and Eberhart [31]. PSO is based on the social behavior of the certain kinds of animals (such as birds and fish) and solves the optimization problem by the cooperation and competition between the particles to search for globally optimal solutions [32]. Each particle adjusts its flying according to its own flying experience and its companions flying experience. Each individual is named as a particle which, in fact, represents a potential solution to a problem. PSO starts with a population of random solutions *particles* in a D-dimension space.

1.6.2 Use of PSO Algorithm for Adjusting the Optimal Gain Matrix (K) in the LQR Controller-Based CSC-STATCOM

PSO is the stochastic search technique that leads a set of population in solution space evolved using the principles of position and velocity updation. With successive updating new generation, a set of updated solutions gradually converges to the real solution. The updated solutions are found from two evolution equations of particles which are given below:

First is original velocity update equation:

$$V_i^{t+1} = w \times V_i^t + \varphi_1 \times r_1(P_i - X_i^t) + \varphi_2 \times r_2(P_g - X_i^t) \tag{1.80}$$

$$X_i^{t+1} = X_i^t + V_i^{t+1} \tag{1.81}$$

where the ith particle is represented by $X_i = (x_{i1}, x_{i2}, \ldots, x_{iD})$. P_i is the best local search position (P_{best}) of the ith particle. P_g is the best global search position (g_{best}) of the ith particle. X_{it} is the previous position and V_{it} is the previous velocity of the ith particle. w is the inertia weight.

At each step in PSO algorithm, every particle velocity is changed according to its P_{best} and g_{best} using Eq. (1.80). This equation has three parts. The first part is represented by previous velocity. The second term is "cognition" part of the particle and "social" part is the third term of the equation. Cognition part represents its personal best experience (position, local search). Social part is also called the group's best experience (global search). New position of the particle is found by Eq. (1.81). In these equations, φ_1 and φ_2 are two positive constants which are recommended as the integer values. These constant values are also called study factor. r_1 and r_2 are the random values in between 0 and 1. These values are used to balance the "cognition" component and "social" component. These parameters determine the performance of PSO, and slightly different settings may lead to very different performance. Value of inertia weight (ω) plays the role of balancing the local search and global search. A large inertia weight is suitable for global search, and a small inertia weight is beneficial for local search. So the value of ω is taken very carefully to make balance in between global and local searching areas.

The fitness function is a particular criterion that is used to evaluate an automatic iterative search in AI techniques such as GA and PSO [33]. So the PSO search algorithm tries to minimize the fitness function through the velocity and position updation process. In this process, the parameters are adjusted to give best value close to the desired response. So fitness function is the main factor in the PSO technique. And the LQR controller is based on the minimization of the cost function (J_C). Therefore in this chapter, the cost function of LQR is used like as a fitness function of the PSO algorithm. This is defined as

$$\text{Min } J_c = \int_0^t x^T(t) \times Q \times x(t)dt + \int_0^t u^T(t) \times R \times u(t)dt \tag{1.82}$$

where t is the simulation time. Min J_C represents the value of minimum cost function. This is found by optimized Q and R matrices. The main objective of PSO technique is minimization of fitness function to find the optimal values of Q and R matrices for controller gain matrix (K_{PSO}). So first 5×5 matrix elements of PSO particles are represented Q matrix, and last 2×2 matrix elements of PSO particles are represented R matrix. Using PSO algorithm, the optimal gain K_{PSO} with the desired eigenstructure in the LQR method can be obtained by choosing the best values of Q and R matrices. These optimal values of controller parameters are found as in Table 1.1. The final configuration of the proposed PSO based LQR controller for the CSC-STATCOM is shown in Fig. 1.15.

The pseudocode of the PSO algorithm for the LQR controller design is provided in Fig. 1.16. To achieve fine-tuning, the parameter setting of PSO is considered very

Table 1.1 Ideal parameters of the LQR controller-based CSC-STATCOM using PSO technique

System parameters	Values
Weighting matrix (Q)	$\begin{bmatrix} 1.54 & 0 & 0 & 0 & 0 \\ 0 & 15.459 & 0 & 0 & 0 \\ 0 & 0 & 100.978 & 0 & 0 \\ 0 & 0 & 0 & 5.333 & 0 \\ 0 & 0 & 0 & 0 & 0.361 \end{bmatrix}$
Weighting matrix (R)	$\begin{bmatrix} 0.434 & 0 \\ 0 & 0.045 \end{bmatrix}$
Ideal gain matrix (K_{GA})	$\begin{bmatrix} -1.8607 & 6.8249 & -0.2477 & 3.8515 & 0.0383 \\ -0.9522 & 1.1156 & 46.0465 & 0.3698 & 5.0417 \end{bmatrix}$

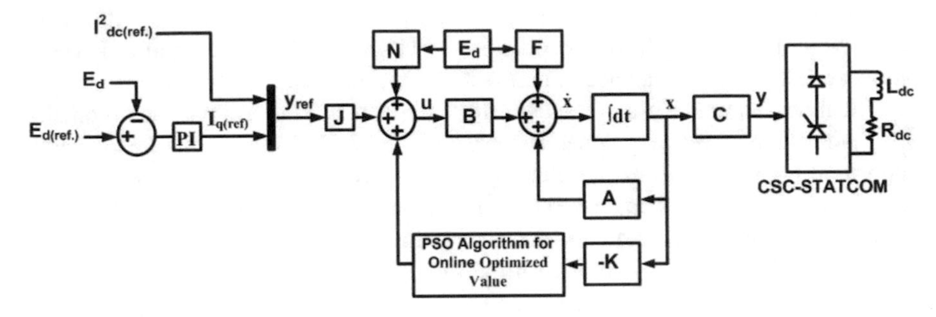

Fig. 1.15 Control structure of the online PSO based controller

```
For each particle
   Initialize particle
END
Do
   For each particle
      Calculate fitness value
      If the fitness value is better than the best fitness value (PBest) in history
         set current value as the new PBest
   End

   Choose the particle with the best fitness value of all the particles as the gBest
   For each particle
      Calculate particle velocity according Equation (80)
      Update particle position according Equation (81)
   End
While maximum iterations or minimum error criteria is not attained
```

Fig. 1.16 Pseudocode of the PSO algorithm

carefully and important. The set of PSO parameters is mentioned in Appendix A, which are used in the simulation studies. The flow chart in Fig. 1.17 explained the main steps in the process of PSO based LQR controller for CSC-STATCOM in brief.

In this flowchart the stopping criterion is the maximum number of iterations or an admissible fitness value for completing the PSO algorithm process. During this operation, the minimization of the objective function is considered with the maximum number of iterations to find the LQR controller parameters. Finally, along with a desired fitness function, an optimum set of the Q and R matrices are obtained through a PSO technique. After completing the steps of PSO algorithm with the assurance of the excellent tuning of controller parameters, the required outcomes for the input-output response of CSC-STATCOM are obtained in terms of less overshoot, less oscillations, low settling time, and little steady-state error

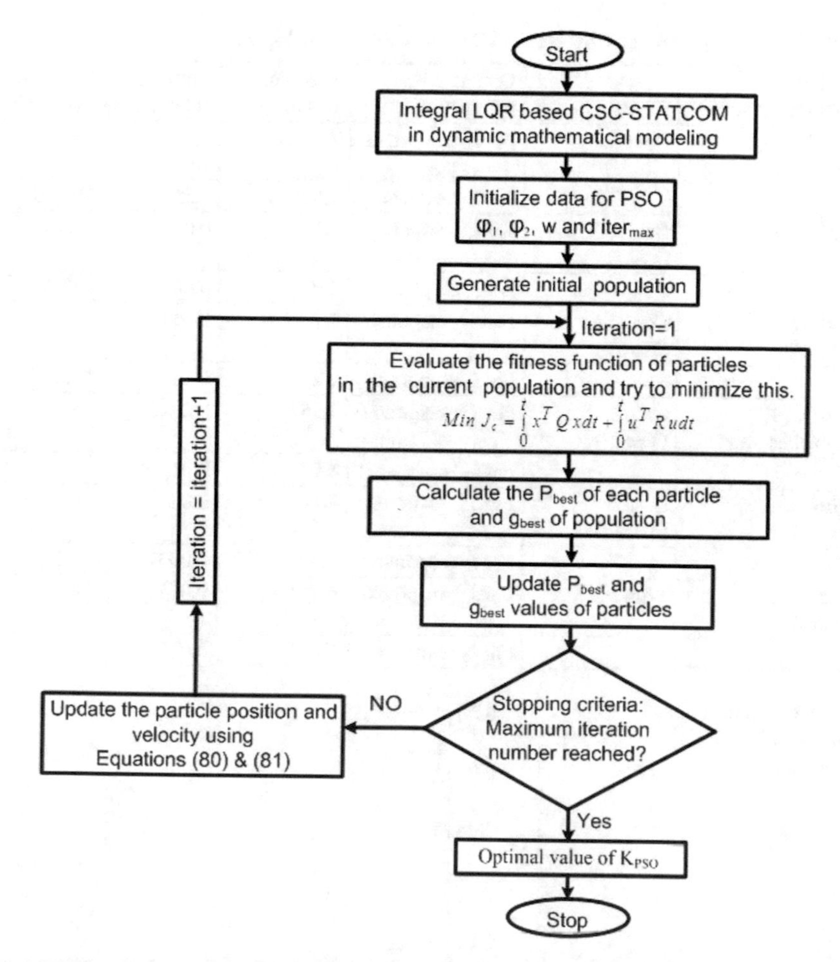

Fig. 1.17 Steps of proposed PSO algorithm for LQR based CSC-STATCOM

which are mentioned in Table 1.2. It can be seen that minimum fitness function (Min J_C) is obtained at seventh iteration in Fig. 1.18. This figure also represented the convergence of PSO.

Now, the searching problem for the desired controller parameters is solved out by PSO based AI technique. In the following section proposed PSO based LQR controller in CSC-STATCOM is considered for the outcomes of input-output. After that, comparison in between different controllers is also shown.

Table 1.2 Outcomes of the CSC-STATCOM with distinct topologies

System with distinct topologies	Case	Output parameters	Rise time (t_r) (ms)	Settling time (t_s) (s)	Overshoot (M_p)%
Without LQR controller	Open loop	For I_{dc}^2 response	2658.5	2.92	33.09
		For i_q response	$8.50*10^{-1}$	1.55	97.06
With LQR controller	Case 1 (TEM1)	For I_{dc}^2 response	3.7	0.01	4.31
		For i_q response	2.0	0.008	10.11
With LQR controller	Case 2 (TEM2)	For I_{dc}^2 response	7.1	0.013	0
		For i_q response	2.2	0.0014	4.6
With LQR controller	Case 3 (TEM3)	For I_{dc}^2 response	16.5	0.0316	0
		For i_q response	8.0	0.0130	0.782
With LQR controller	Case 4 (TEM4)	For I_{dc}^2 response	20	0.035	0
		For i_q response	2.1	0.0032	2.21
With LQR controller	With PSO	For I_{dc}^2 response	6.7	0.0121	0
		For i_q response	0.4	0.0004	0.39

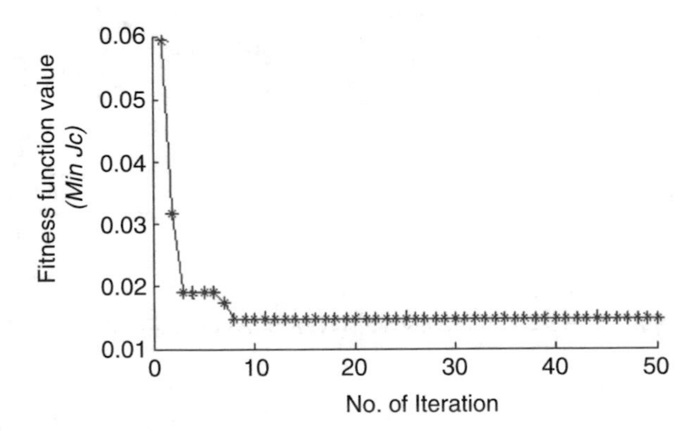

Fig. 1.18 Convergence behavior of the PSO algorithm

1.7 Performance of Optimal LQR Controller-Based CSC-STATCOM Using AI Technique

In this section, the impact of the introduced topology has been observed for the desired criterion of the test system through MATLAB/Simulink. The matrix elements of Q and R are used as mentioned in Table 1.1. These values are obtained by the proposed PSO based LQR controller. The simulation results of the PSO based controllers with CSC-STATCOM are shown in Figs. 1.19, 1.20, and 1.21. From these figures, it is observed that PSO technique-based LQR controller gives better results than the trial and error method (TEM)-based LQR controller. Clearly,

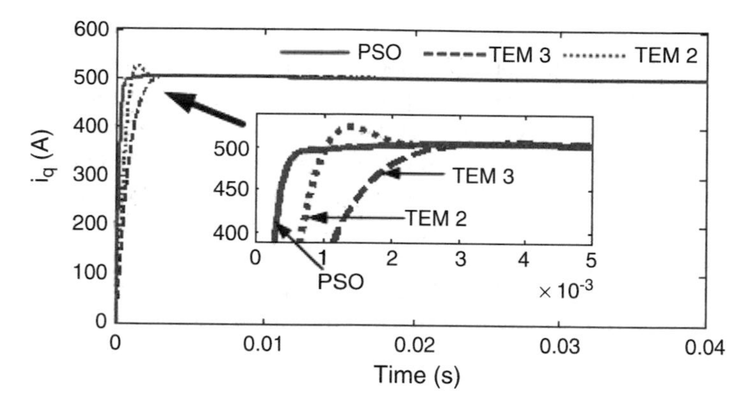

Fig. 1.19 i_q response with LQR controller based on PSO algorithm

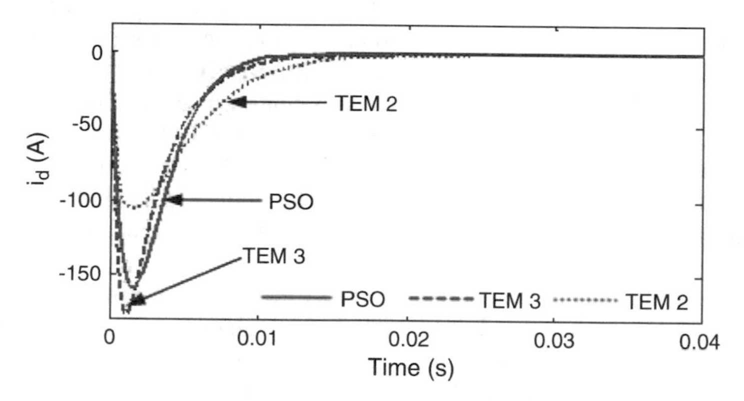

Fig. 1.20 i_d response with LQR controller based on PSO algorithm

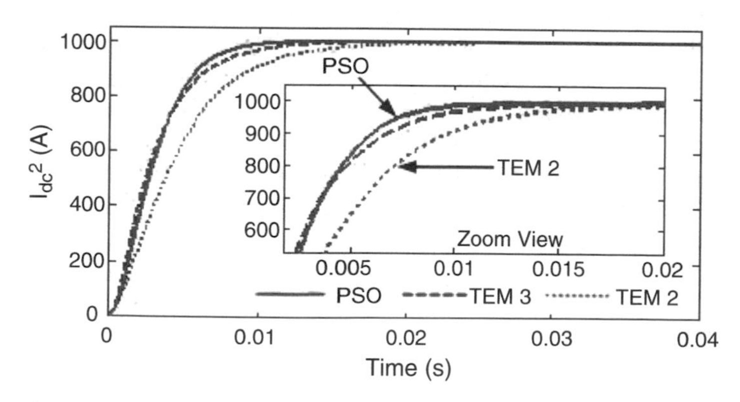

Fig. 1.21 I_{dc}^2 response with LQR controller based on PSO algorithm

22. Ali N, Amin V (2007) An LQR/Pole placement controller design for STATCOM. In: Proceedings of the Conference Record IEEE CCC, July 2007, pp 189–193
23. Ni Y, Jiao L, Chen S, Zhang B (1998) Application of a nonlinear PID controller on STATCOM with a differential tracker. In: International Conference on Energy Management and Power Delivery, EMPD-98, New York, pp 29–34
24. Rao P, Crow ML, Yang ZP (2000) STATCOM control for power system voltage control applications. IEEE Trans Power Deliv 15(4):1311–1317
25. Bilgin HF, Ermis M, Kose KN, Cetin A, Cadirci I, Acik A, Demirci T, Terciyanli A, Kocak C, Yorukoglu M (2007) Reactive-power compensation of coal mining excavators by using a new-generation STATCOM. IEEE Trans Ind Appl 43(1):97–110
26. Gopal M (1993) Modern control system theory. New Age International, New Delhi
27. Zuliari EA, Abdillah M, Rosyadi M (2009) Design optimal linear quadratic regulator for static synchronous compensator (Statcom) using particle swarm optimization. In: Conference: Proceeding of the 1st National Seminar on Applied Technology, Science, and Arts. At Institut Teknologi Sepuluh Nopember (ITS), Surabaya
28. Tewari A (2002) Modern control design with MATLAB and SIMULINK. Wiley. ISBN:978-0-471-49679-3
29. Ogata K (2010) Modern control engineering, 5th edn. Prentice Hall, Upper Saddle River
30. Warwick K, Ekwue A, Aggarwal R (1997) Artificial intelligence techniques in power systems. Institution of Electrical Engineers, London
31. Shi YH, Eberhart RC (1998) A modified particle swarm optimizer. In: Proceeding of IEEE International Conference on Evolutionary Computation (ICEC'98), Anchorage, May 1998, pp 69–73
32. Rini DP, Shamsuddin SM, Yuhaniz SS (2011) Particle swarm optimization: technique, system and challenges. Int J Comput Appl 14(1):19–27. (ISSN:0975–8887)
33. Engelbrecht AP (2007) Computational intelligence: an introduction, 2nd edn. Wiley, The Atrium/Southern Gate/Chichester/West Sussex

Chapter 2
A Brief Review and Comparative Study of Nature-Inspired Optimization Algorithms Applied to Power System Control

Nour E. L. Yakine Kouba and Mohamed Boudour

2.1 Introduction

During several years, remarkable efforts have been established for enhancing power system control and overcoming frequency and voltage instability challenges. Many reported researches in the field of power system control design have been proposed. Mainly, the auxiliary controller in power system called as the Automatic Generation Control (AGC) includes two kinds of regulation loops, the automatic voltage regulator (AVR) and the load frequency control (LFC), which rely on the excitation system and the speed regulation system [1]. These systems take an important part of the generator and play a vital role in maintaining the stability of the power system [2–3].

In 90% of industrial systems, the ancillary services are using a PI or PID controller as a major control agent to deal with system disturbances [4]. These controllers are successfully used in many real-life applications. In recent decades, various studies have proposed a new control concept with the use of the artificial intelligence techniques [5–7]. These methods received a remarkable appreciation in several research areas, especially for parameters optimization [8–10]. The optimal control design issues for enhancing network performance in power systems have attracted great attention in recent years. Many researchers have deliberated the application of nature-inspired and bio-inspired optimization methodologies to improve power system control actions using optimal PID controller [11–14].

Currently, the research world recognize a fast development in the optimization algorithms that mimic the social, living, hunting, and feeding mechanism of many kinds of creatures in nature such as wolves, elephants, whale, fishes, spider, bees,

N. E. L. Y. Kouba (✉) · M. Boudour
Laboratory of Electrical and Industrial Systems, Faculty of Electrical Engineering and Computing, University of Sciences and Technology Houari Boumediene, Algiers, Algeria

X. Li, K.-C. Wong (eds.), *Natural Computing for Unsupervised Learning*, Unsupervised and Semi-Supervised Learning,
https://doi.org/10.1007/978-3-319-98566-4_2

ants, bats, and birds. In reference [15], Dr. X.S Yang has proposed a very interesting literature review about metaheuristics algorithms in the past few years. In [16], Deepthi S and Aswathy Ravikumar have proposed a study from the perspective of nature-inspired metaheuristic optimization algorithms. In [17], Iztok Fister Jr. et al. have proposed *A Brief Review of Nature-Inspired Algorithms for Optimization*. In [18], Michael A. Lones has discussed several metaheuristic algorithms.

According to the enormous number of the recently developed metaheuristic algorithms, it is very difficult to cover most powerful metaheuristic algorithms. As claimed in many published papers, some algorithms have shown very powerful performances and are routinely used in solving many issues. However, with the intention to help interested readers about metaheuristics to adapt a good optimization algorithm with their application, the choice of this algorithm is a difficult task.

In this view, and because we can't cover all metaheuristics algorithms, we have tried through this work to focus on the most popular metaheuristics and give a brief review list of all efficient optimization tools published in the literature during the last 4 years, so as to inspire further research. Among the newly developed optimization strategies published between 2014 and 2017 [19–30], the most popular that have proved to be very efficient, we can site:

Grey Wolf Optimizer (GWO) GWO is a novel nature-inspired optimization tool that mimics the leadership hierarchy and hunting mechanism of gray wolves in nature. GWO was developed by Mirjalili et al. in 2014 [19].

Ant Lion Optimizer (ALO) ALO is a nature-inspired algorithm proposed recently by Seyedali Mirjalili in 2014. ALO algorithm mimics the hunting mechanism of a special kind of ant in nature called antlions [20].

Multi-Verse Optimizer (MVO) MVO is a novel nature-inspired algorithm inspired from three concepts in cosmology: white hole, black hole, and wormhole. The MVO algorithm was proposed by Seyedali Mirjalili et al. in 2014 [21].

Dragonfly Algorithm (DA) The DA algorithm was proposed by Seyedali Mirjalili in 2015. The main inspiration of the DA comes from the static and dynamic swarming mechanism of dragonflies in nature [22].

Moth-Flame Optimization Algorithm (MFOA) MFOA is a novel nature-inspired optimization tool proposed by Seyedali Mirjalili in 2015. The main inspiration of MFOA optimizer is the transverse orientation method, which is the navigation method of moths in nature [23].

Whale Optimization Algorithm (WOA) WOA is a new optimization strategy proposed by Seyedali Mirjalili and Andrew Lewis in 2015. The WOA mimics the social behavior and the bubble-net hunting mechanism of humpback whales in oceans [24].

Elephant Herding Optimization (EHO) The EHO algorithm was proposed by Gai-Ge Wang et al. in 2015. This algorithm was introduced for the first time in the third International Symposium on Computational and Business Intelligence (2015).

EHO is considered as a new kind of swarm-based metaheuristic search algorithm, which is inspired by the herding behavior of elephant group nature [25].

Lion Optimization Algorithm (LOA) This algorithm was introduced by Maziar Yazdani and Fariborz Jolai in 2015. LOA mimics the special lifestyle cooperation characteristics of lions in nature [26].

Monarch Butterfly Optimization (MBO) MBO is a new optimization algorithm developed in 2015 by Gai-Ge Wang et al. MBO mimics the migration behavior of monarch butterfly during the late summer/autumn from the Northern USA and Southern Canada to Mexico [27].

Grasshopper Optimization Algorithm (GOA) GOA was developed in 2016 by Shahrzad Saremi et al. This new kind of optimization algorithm mimics the behavior of grasshopper swarms in nature [28].

Salp Swarm Algorithm (SSA) SSA is a novel bio-inspired optimizer proposed by Seyedali Mirjalili et al. in 2016. The SSA was inspired from the swarming behavior of salps when navigating and foraging in oceans [29].

Electro-Search Algorithm (ESA) ESA is a novel optimization algorithm proposed by Amir Tabari and Arshad Ahmad in 2017. The main inspiration of ESA comes from the movement of electrons through the orbits around the nucleus of an atom. ESA is based on some physical principals such as Bohr model and Rydberg formula [30].

In this work, an optimal solution for the frequency and voltage regulation problem using optimal PID controller is introduced to reduce system fluctuations using a recently developed nature-inspired and bio-inspired algorithms. In this context, a brief literature review on the newly developed optimization algorithms during the last 4 years was performed. The proposed algorithms were used to solve the optimal tuning parameters problem including both of PID and PSS controllers. Finally, a comparative study between these algorithms was carried out.

2.2 Power System Frequency and Voltage Control

In power system, a proper control system that guarantees fast and safe operation for a generation power plant is necessary to ensure a good quality of energy service toward the customer. Frequency and voltage are the most important factors in power system. The process design of both frequency and voltage control must be carefully selected to avoid equipment damage due to failures. Power system control and operation are carried out on varying time scales ranging from seconds to minutes in the case of voltage and frequency control and from minutes to hours in the case of electricity market including intra- and day-ahead power dispatch [31].

Power system control consists of a number of nested control loops that regulate different quantities such as frequency, voltage, and interline power exchange; this

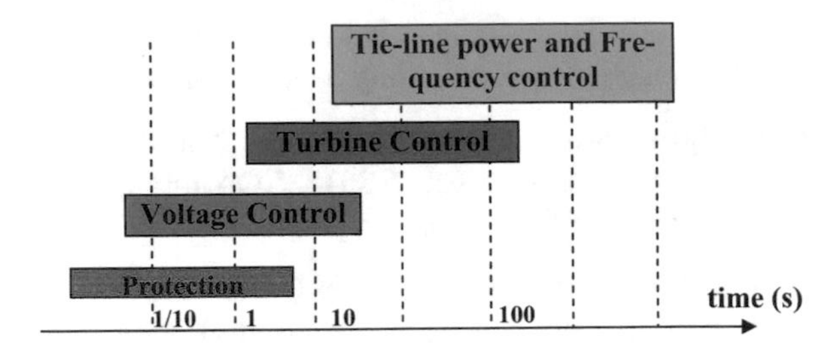

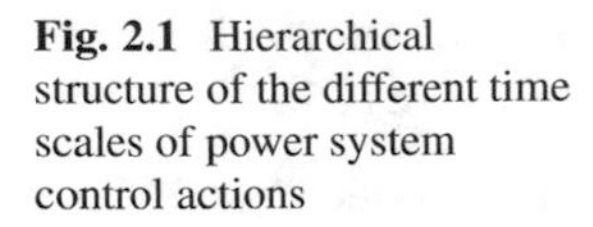
Fig. 2.1 Hierarchical structure of the different time scales of power system control actions

control system is generally known as the Automatic Generation Control (AGC). The hierarchical structure of the different time scales of the associated control system is shown in Fig. 2.1 [32–33].

As shown in Fig. 2.1, the control loops on lower system levels represent the local regulation and operation actions in the generator, which are characterized by smaller time constants such as the protection scheme and the automatic voltage regulator (AVR). In addition, the local speed regulation via the turbine control needs a different time scale, which is characterized by higher time constants compared to the protection and voltage regulation loop. On the other hand, the auxiliary control loops through the AGC system operates in a time scale of tens of seconds to minutes for both of secondary voltage control and secondary frequency control (LFC). Moreover, it is important to mention that both frequency and voltage control are virtually decoupled because they are operated in different time scales. However, the combined model of LFC and AVR creates a mutual interaction between frequency and voltage regulation loops [34].

In this part, a fundamental basic model of a synchronous generator was presented. The used synchronous machine model is shown in Fig. 2.2 [35]. The presented model comprises both of LFC and AVR loops. The LFC loop is used to adjust operating point reference of the selected unit in the aim to maintain the frequency at the scheduled value, where the AVR loop is used for voltage magnitude regulation automatically. LFC and AVR loops are also called active power and reactive power control loop, respectively. The LFC is achieved by the speed control mechanism, where the AVR is achieved by the excitation mechanism, which can control the rotation speed and the field current of the synchronous machine as shown in Fig. 2.2. This study was performed using and extending the linearized AGC system [36].

The block diagram of the used synchronous generator is shown in Fig. 2.3. A simplified frequency regulation system with primary and secondary is shown in Fig. 2.4. A PID controller was added to the LFC loop to enhance the dynamic frequency response [37]. A typical IEEE DC1 AVR loop with excitation system is shown in Fig. 2.5 [38–39]. A PID controller was implemented in the AVR system to keep the voltage magnitude within acceptable limits.

In addition, a Power System Stabilizer (PSS) was added as a supplementary input to the AVR system to mitigate the inter-area oscillations sufficiently in power system as shown in Fig. 2.5 [40–41]. The objective of the PSS system is to add damping

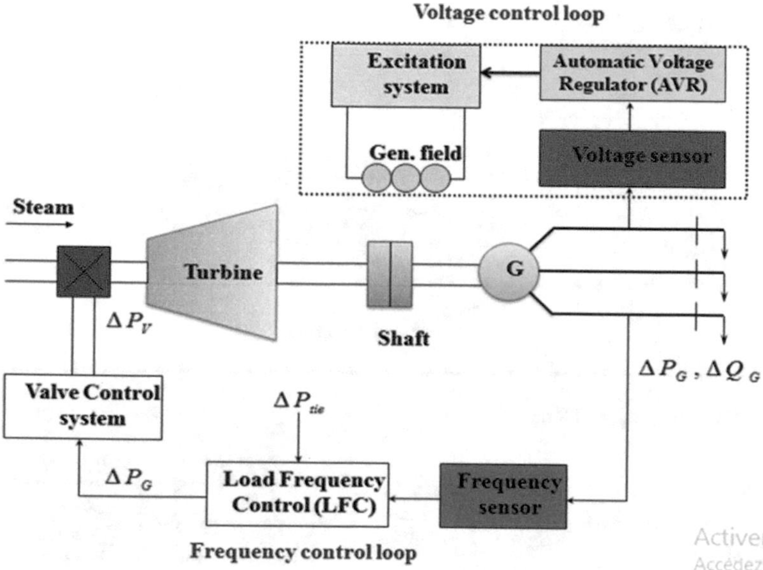

Voltage control loop
Excitation system
Automatic Voltage Regulator (AVR)
Gen. field
Voltage sensor
Steam
Turbine
G
ΔP_V
Shaft
ΔP_G , ΔQ_G
Valve Control system
ΔP_{tie}
ΔP_G
Load Frequency Control (LFC)
Frequency sensor
Frequency control loop

Fig. 2.2 Diagram of LFC and AVR control loops of a synchronous generator

ΔP_L
ΔPm
ΔPe
$\frac{1}{M.s + D}$
Δw
$\frac{1}{s}$
$\Delta\delta$
Ps
K5
K2
K4
Vt
E'_q
$\frac{KG}{Tg.s + 1}$
K6
Efd

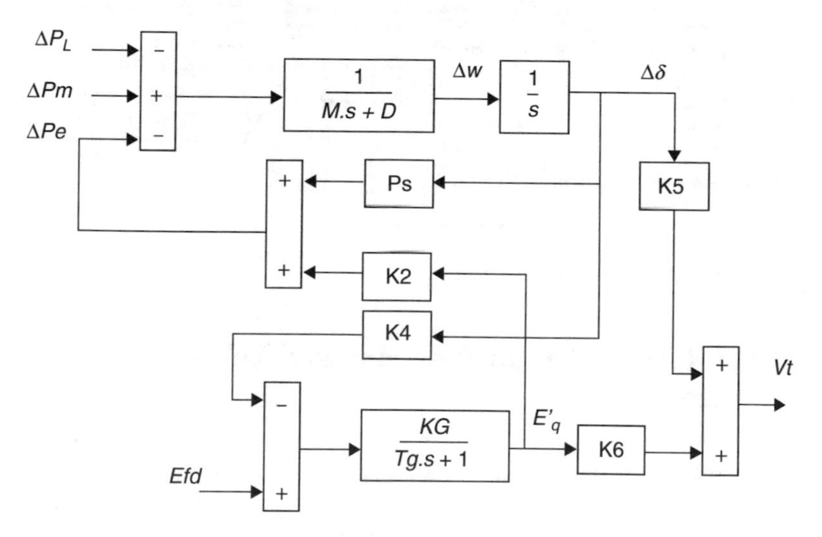

Fig. 2.3 Block diagram of a synchronous generator

to the generator rotor oscillations by controlling its excitation using auxiliary stabilizing signals: (1) frequency (Δf), (2) rotor angle ($\Delta\delta$), and (3) electric power (ΔPe) [42].

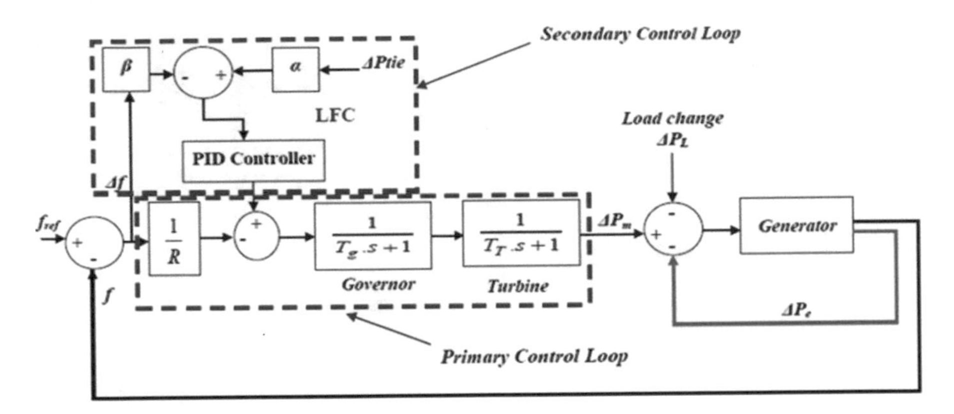

Fig. 2.4 Frequency control loops

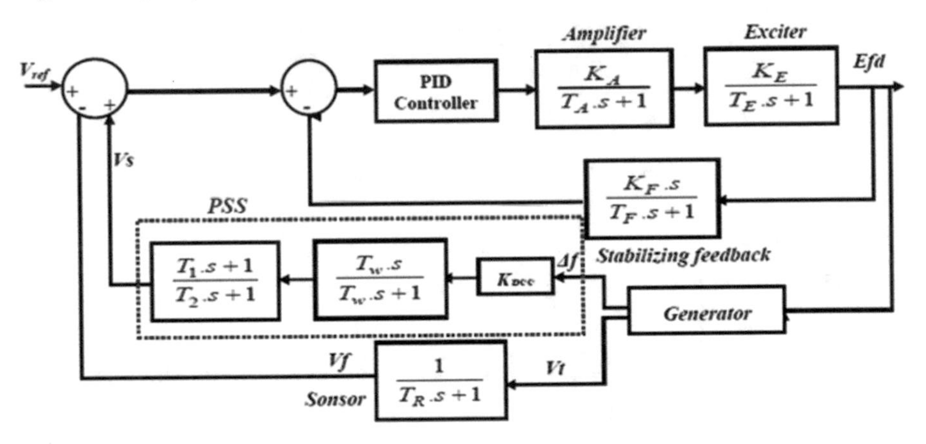

Fig. 2.5 Voltage control loops

2.3 Literature Review on Frequency and Voltage Control Strategies

Nowadays, more and more modern metaheuristic algorithms are developed, and they become increasingly popular. These algorithms are considered to be the most efficient strategies that are used for solving various optimization problems such as the control scope in electrical engineering. This section attempts to summarize the findings of a large number of research works concerning the application of optimization algorithms to power system frequency and voltage control [43–48].

From the literature survey, many researches have been investigated in the scope of power system control and operation. Even with the good intention and effort of many researchers to give the best control scheme and overcome power system instability challenges, there is still a large work to do in the field of optimization to further enhance the power system control performance. It's very difficult to cover all contributions in the scope of optimization techniques yield to the AGC system including LFC and AVR systems with PID and PSS controllers. It should be noted first that each of AGC, LFC, AVR, and AVR-PSS system can be analyzed individually, where there are a lot of published papers that focus on the study of each system separately. Moreover, several papers have discussed the optimization of PID controller gains to improve the LFC loop. On the other hand, there are also a lot of papers that have proposed the optimal tuning of PSS system parameters to overcome some voltage control challenges. Actually, several efforts have been placed on various optimization control schemes for frequency, voltage, and stabilization issues [49–52].

From the past and current research, it can be expected that the future research will focus more and more on the computational optimization techniques that are inspired from nature. In the present work, we have tried to analyze the state of the art of power system control with some applied bio-inspired and nature-inspired optimization algorithms that have been developed and used for LFC and AVR systems through the past few years. This work can be considered as an effort that could help in some way the future researchers that are interested to use metaheuristics optimization algorithms to solve power system stability and control problems. Table 2.1 reports some control strategies and optimization algorithms related to the LFC and AVR systems-based I, PI, PID, and PSS controllers that have been published during the last 5 years. Due to the constraint of paper length, we only cited the most popular applied strategies, where there are still a lot of works that should be included.

2.4 Simulation Results and Discussion

In this part, the dynamic responses of frequency, voltage, and rotor angle of the simulated single-area power system are presented. A comparative analysis between the applied optimization algorithms is performed as shown in Figs. 2.6, 2.7, and 2.8. The proposed single-area power system was simulated during 0.1 pu load change. The simulation data are cited in Tables 2.2, 2.3, and 2.4, where Tables 2.5 and 2.6 provide the values of the PID controller gains and the PSS controller parameters that are tuned using seven optimization algorithms: GWO, ALO, MVO, MBO, WOA, PSO, and GA. The results are compared in view of settling time, peak overshoots, and peak undershoots as presented in Figs. 2.6, 2.7, and 2.8. The optimization problem is formulated using the same objective function, which is the integral time multiplied by absolute error (*ITAE*) as given in Eq. (2.1):

$$\text{ITAE} = \int_{0}^{\text{tsim}} t.\left(\left|\Delta f\right| + \left|\Delta V\right|\right).dt \qquad (2.1)$$

Table 2.1 Literature review on some recent LFC and AVR control strategies

Authors	Proposed work	Year
S. R. Manickam and M. T. A. Mustafa	Optimal LFC using bat algorithm [53]	2014
Binod Kumar Sahu et al.	Optimal AGC based on Hybrid Differential Evolution-Particle Swarm Optimization [54]	2014
S.Satyanarayana et al.	Mutual effect between LFC and AVR loops [2]	2014
E.S. Ali	Optimization of PSS using bat search algorithm [55]	2014
O. Bendjeghaba and I. B. Saida	Bat algorithm for optimal tuning of PID controller in an AVR system [56]	2014
Banaja M et al.	Differential Evolution Algorithm applied to LFC system [57]	2014
Rabindra Kumar Sahu et al.	Optimal AGC using Gravitational Search Algorithm [58]	2014
K. Jagatheesan et al.	Artificial intelligence for LFC system [6]	2015
G. G. Samuel and C. C. A. Rajan.	Hybrid PSO-GA and PSO-SFLA for long-term generator maintenance scheduling [59]	2015
Rabindra Kumar Sahu et al.	A hybrid Firefly Algorithm and pattern search technique for AGC system [60]	2015
Rabindra Kumar Sahu et al.	Hybrid PSO-PS optimized fuzzy PI controller for AGC system [61]	2015
B. K. Sahu et al.	Teaching-learning based optimization algorithm based fuzzy-PID controller for AGC system [62]	2015
Dhanraj chitara et al.	Optimal tuning of PSS using Cuckoo Search Algorithm [63]	2015
İ. Eke et al.	Robust tuning of Power System Stabilizer by using orthogonal learning artificial bee colony [64]	2015
Mahmut T. et al.	Tuning of optimal classical and fractional order PID for AGC using bacterial swarm optimization [65]	2015
V Soni et al.	Hybrid Grey wolf optimization-pattern search-optimized 2dof-Pid controllers for LFC system [66]	2016
Esha Gupta and Akash Saxena	Antlion optimizer-based regulator In Automatic Generation Control of Interconnected Power System [67]	2016
Bestoun S. Ahmed et al.	Optimum design of $PI\lambda D\mu$ controller for an AVR system using combinatorial test design [68]	2016
Salma KESKES et al.	Optimal tuning of PSS using genetic algorithm	2017
M. Venkatesh and G. Sudheer	Optimal LFC using Dragonfly Algorithm [69]	2017
Santigopal pain and Parimal Acharjee	Optimal LFC system using chaotic exponential PSO algorithm [70]	2017
Amit Kumar and Sathans Suhag	Multiverse optimized fuzzy-PID for LFC system [71]	2017
DK Sambariya and Rajendra Fagna	Elephant Herding Optimization-based PID controller design for LFC in power system [72]	2017
Prasenjit Dey	Tuning of PSS using collective decision optimization algorithm [73]	2018
Rahmat Khezri et al.	LFC-based SCA-optimized FOPID controllers [74]	2018
Piyush Mathur et al.	Comparative analysis of PID tuning of AVR [75]	2018

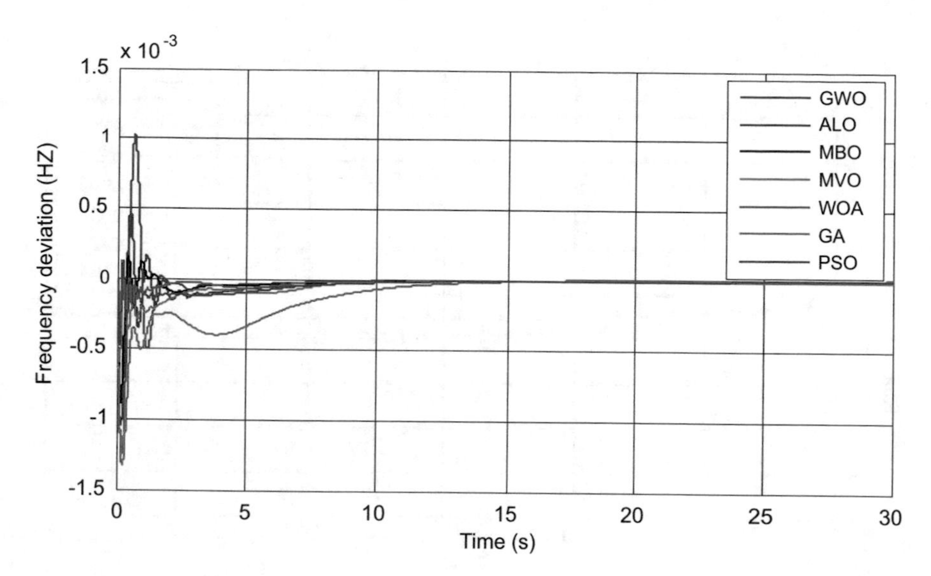

Fig. 2.6 Frequency deviation

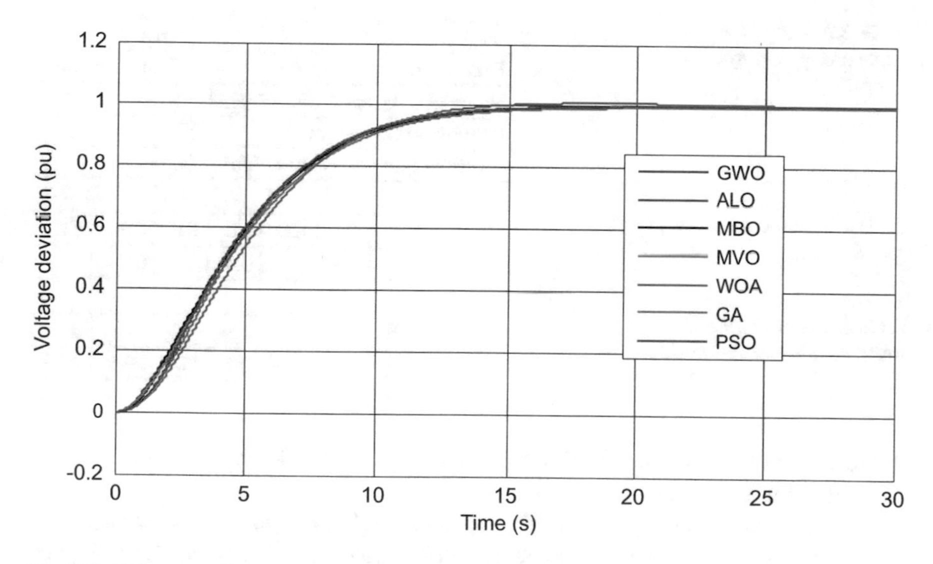

Fig. 2.7 Voltage profile

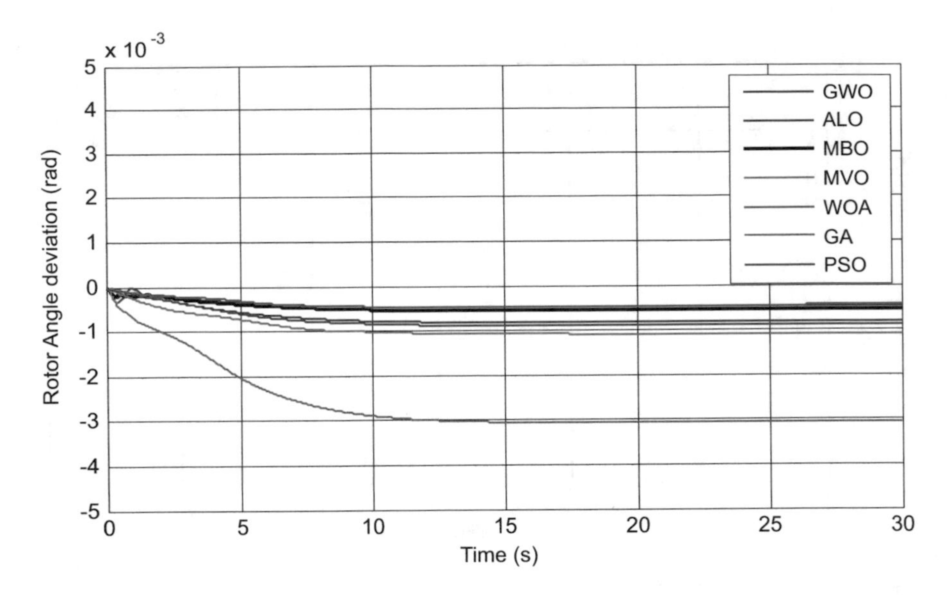

Fig. 2.8 Rotor angle deviation

Table 2.2 Data of the single-area power system

	Area
Parameters	Equivalent area
Nominal frequency value f [HZ]	50
Inertia constant M [s]	10
Load-damping constant D [pu]	0.9

Table 2.3 Data of the AVR system

KA	TA	KE	TE	KF	TF	TR
10	0.1	1	0.4	2	0.04	0.05

Table 2.4 Data of the governor-turbine/LFC system

K_{Gh}	T_g	T_T	β
20	0.2	0.5	20.6

where Δf presents frequency deviation, ΔV presents voltage deviation, and $\Delta\delta$ presents rotor angle deviation.

The dynamic performances of the proposed single-area power system including frequency, voltage, and rotor angle are compared using different optimization algorithms. The results are obtained by considering 0.1 pu step load disturbance. Frequency, voltage, and rotor angle deviations (Δf, ΔV, $\Delta\delta$) presented in Figs. 2.6, 2.7, and 2.8 show the advantage of using metaheuristics algorithms in parameters optimization issue. The presented results show clearly the effect of the optimal PID and PSS controllers in maintaining system stability in case of load change. It can be seen that the fluctuations of system are suppressed most effectively when using optimal controller. Most of the applied algorithms show the same performance in view of minimizing settling time, peak overshoots, and peak undershoots.

Table 2.5 Optimal PID controller gains

Methods		Parameters		
		K_P	K_I	K_D
GWO	LFC	10.5347	50	3.5905
	AVR	7.4235	48.2365	2.3648
ALO	LFC	17.7351	29.2154	4.9520
	AVR	17.3750	24.0742	11.9641
MBO	LFC	13.4017	43.5384	7.3248
	AVR	13.2125	47.4215	7.3023
MVO	LFC	8.68399	22.1971	16.3107
	AVR	40.2089	0.43727	4.90523
WOA	LFC	17.9962	50	31.0871
	AVR	22.1028	2.1142	12.5465
PSO	LFC	23.1472	26.8743	24.1479
	AVR	27.8710	30	17.1840
GA	LFC	5.0471	7.8386	3.6137
	AVR	7.3540	3.2357	3.8642

Table 2.6 Optimal PSS controller parameters

Methods	Parameters			
	K_{Pss}	T_w	T_1	T_2
GWO	0.1313	0.1367	0.2128	1.0292
ALO	4.0847	6.1429	0.6487	1.8714
MBO	2.7421	3.8391	0.1957	1.1464
MVO	10.5207	7.15521	0.6107	4.6859
WOA	1.9994	4.9998	0.1905	1.1429
PSO	1.9829	5.4280	1.8749	2.2198
GA	2.4251	11.3759	0.9427	2.7428

It is clear that some optimization techniques show more oscillations, where others are very effective and fast in system oscillations damping, which is due to the control parameters of each method. In addition, the simulation time and number of iteration can change from method to other. However, the same objective is achieved using these methods, which is, using those algorithms, the PID gains and the PSS parameters are optimized very perfectly, which lead to a significant improvement in the LFC and AVR loops.

It can be concluded from the presented results in this work that the implementation of nature-inspired optimization algorithms to design optimal controller improves effectively the performance of the electric network in terms of minimum over−/undershoot and lesser settling time.

2.5 Conclusion

The current work investigates on the application of some nature-inspired meta-heuristics algorithms for enhancing power system frequency and voltage control. A brief introduction on the recently optimization techniques has been proposed to enrich the literature review of optimization scope and to shed some light on this research area for readers unfamiliar with those techniques. Then, this work has put emphasis on the literature review of AGC system based on some published control strategies using intelligent and optimization algorithms during the past decades that have been applied on LFC and AVR systems such as ANN, logic flow, GA, PSO, ABC, ACO, etc. In addition, an application of some newly developed nature-inspired optimization algorithms such as GWO, ALO, MVO, MBO, and WOA for optimal tuning of PID controller and Power System Stabilizer (PSS) parameters has been proposed with a comparative study through the popular PSO and GA algorithm. The main aim was to enhance frequency and voltage regulation during disturbances. The chosen optimization strategies have been tested on a typical single-area with standard LFC and AVR controller.

A time domain simulation was carried out during load changes. The transient response of frequency and voltage variations has been analyzed. Furthermore, a comparative analysis was performed between the used optimization algorithms.

The obtained results are compared in view of settling time, peak overshoots, and peak undershoots. The results have shown the good contribution of the optimized PID and PSS controllers in improving power system regulation. Moreover, the results have proven the advantage and effectiveness of the used optimization algorithms for parameters optimization and adjustment.

In a future work, the proposed optimization algorithm could be used also to solve parameter estimation problem in power system, such as FACTs, HVDC link, and wind farm, or for storage system sizing or even for multi-agent system design. In addition, the application could also be extended to multi-area interconnected or deregulated power system with integration of renewable energy resource and considering nonlinearity constraints and electricity market scenario.

References

1. Elgerd OI (1981) Control of electric power systems. IEEE Control Syst Mag 1(2):4–16
2. Satyanarayana S, Sharma RK, Mukta (2014) Mutual effect between LFC and AVR loops in power plant. Electr Electron Eng Int J 3(1):61–69
3. Anbarasi S, Muralidharan S (2014) Transient stability improvement of lfc and avr using bacteria foraging optimization algorithm. Int J Innov Res Sci Eng Technol 3(3):124–129
4. Marzoughi A et al (2012) Optimized proportional integral derivative (PID) controller for the exhaust temperature control of a gas turbine system using particle swarm optimization. Int J Phys Sci 7(5):720–729
5. Bahgaat NK, Moustafa Hassan MA (2016) Swarm intelligence PID controller tuning for AVR system. In: Advances in Chaos theory and intelligent control. Springer, Cham, pp 791–804

6. Jagatheesan K, Dey N, Anand B, Ashour AS (2015) Artificial intelligence in performance analysis of load frequency control in thermal-wind-hydro power systems. Int J Adv Comput Sci Appl 6(7):203–212
7. Dahiya P et al (2016) Comparative performance investigation of optimal controller for AGC of electric power generating systems. Automatika 57(4):902–921
8. Yang X-S (2011) Review of meta-heuristics and generalised evolutionary walk algorithm. Int J Bio-Insp Comp 3(2):77–84
9. Zhang LD, Jia L, Zhu WX (2012) A review of some intelligent optimization algorithms applied to intelligent transportation system. Adv Mater Res 383. Trans Tech Publications
10. Yang X-S, Koziel S, Leifsson L (2014) Computational optimization, modelling and simulation: past, present and future. Procedia Comput Sci 29:754–758
11. Kouba NELY et al (2017) A new optimal load frequency control based on hybrid genetic algorithm and particle swarm optimization. Int J Electr Eng Inform 9(3):418–440
12. El Yakine Kouba N, Menaa M, Hasni M, Boudour M (2016) A novel optimal frequency control strategy for an isolated wind–diesel hybrid system with energy storage devices., SAGE. Wind Eng 40(6):497–517
13. Rahimi K, Famouri, P (2013) Performance enhancement of automatic generation control for a multi-area power system in the presence of communication delay. North American Power Symposium (NAPS), IEEE
14. Oonsivilai R, Oonsivilai A (2011) Gas turbine optimal PID tuning by genetic algorithm using MSE. World Acad Sci Eng Technol 5(12):257–262
15. Yang X-S (2010) Nature-inspired metaheuristic algorithms. Luniver press, UK
16. Deepthi S, Ravikumar A (2015) A study from the perspective of nature-inspired metaheuristic optimization algorithms. Int J Comput Appl 113:9
17. Fister I, et al (2013) A brief review of nature-inspired algorithms for optimization. arXiv preprint arXiv:1307.4186
18. Lones MA (2014) Metaheuristics in nature-inspired algorithms. Proceedings of the companion publication of the 2014 annual conference on genetic and evolutionary computation, ACM
19. Mirjalili S, Mirjalili SM, Lewis A (2014) Grey wolf optimizer. Adv Eng Softw 69:46–61
20. Mirjalili S (2015) The ant lion optimizer. Adv Eng Softw 83:80–98
21. Mirjalili S, Mirjalili SM, Hatamlou A (2016) Multi-verse optimizer: a nature-inspired algorithm for global optimization. Neural Comput Applic 27(2):495–513
22. Mirjalili S (2016) Dragonfly algorithm: a new meta-heuristic optimization technique for solving single-objective, discrete, and multi-objective problems. Neural Comput Appl 27(4):1053–1073
23. Mirjalili S (2015) Moth-flame optimization algorithm: a novel nature-inspired heuristic paradigm. Knowl Based Syst 89:228–249
24. Mirjalili S, Lewis A (2016) The whale optimization algorithm. Adv Eng Softw 95:51–67
25. Wang GG, Deb S, Leandro dos SC (2015) Elephant herding optimization. Computational and Business Intelligence (ISCBI), 2015 IEEE 3rd International Symposium, 1–5
26. Yazdani M, Jolai F (2016) Lion optimization algorithm (LOA): a nature-inspired metaheuristic algorithm. J Comput Des Eng 3(1):24–36
27. Wang G-G, Deb S, Cui Z (2015) Monarch butterfly optimization. Neural Comput Applic 1–20. https://doi.org/10.1007/s00521-015-1923-y
28. Saremi S, Mirjalili S, Lewis A (2017) Grasshopper optimisation algorithm: theory and application. Adv Eng Softw 105:30–47
29. Mirjalili S et al (2017) Salp swarm algorithm: a bio-inspired optimizer for engineering design problems. Adv Eng Softw 114:163–191
30. Tabari A, Ahmad A (2017) A new optimization method: electro-search algorithm. Comput Chem Eng 103:1–11
31. Ulbig A et al (2011) Framework for multiple time-scale cascaded MPC application in power systems. IFAC Proc 44(1):10472–10480
32. Hasan N et al (1978) Automatic generation control problem in interconnected power systems. Proc Inst Electr Eng 125:385

33. Abdulraheem BS, Gan CK (2016) Power system frequency stability and control: survey. Int J Appl Eng Res 11(8):5688–5695
34. Kouba NELY, Menaa, M, Hasni, M, Boudour, M (2015) Optimal control of frequency and voltage variations using PID controller based on particle swarm optimization, IEEE, 4th international conference on systems and control ICSC'2015, Sousse, Tunisia, pp 424–429
35. Demello FP, Concordia C (1969) Concepts of synchronous machine stability as affected by excitation control. IEEE Trans Power Syst 88(4):316–329
36. Rakhshani E, Sadeh J (2010) Application of power system stabilizer in a combined model of LFC and AVR loops to enhance system stability. International Conference on IEEE Power System Technology (POWERCON)
37. Soundarrajan A, Sumathi S, Sundar C (2010) Particle swarm optimization based LFC and AVR of autonomous power generating system. Int J Comput Sci 37(1):1–8
38. Report IEEE (1968) Computer representation of excitation systems. IEEE Trans Power Syst 6:1460–1464
39. Faiz J, Shahgholian GH, Arezoomand M (2007) Analysis and simulation of the AVR system and parameters variation effects. In: International conference on IEEEPower engineering, energy and electrical drives, POWERENG 2007, pp 450–453
40. Chatterjee A, Ghoshal SP, Mukherjee V (2010) Transient performance improvement of thermal system connected to grid using distributed generation and capacitive energy storage unit. Int J Electr Eng Inform 2(3):159
41. Report, IEEE (1981) Excitation system models for power system stability studies. IEEE Trans Power Syst 2:494–509
42. De Mello FP, Hannett LN, Undrill JM (1978) Practical approaches to supplementary stabilizing from accelerating power. IEEE Trans Power Syst 5:1515–1522
43. Pandey SK, Mohanty SR, Kishor N (2013) A literature survey on load frequency control for conventional and distribution generation power systems. Renew Sust Energ Rev 25:318–334
44. Kouba NELY, Menaa M , Hasni M, Boudour M (2017) Computational intelligence applied to power system restoration: a case study of monarch butterfly optimization algorithm. ICATS, 11th -12th Dec 2017, Annaba, p 1–6
45. Urfi M, Prashar S (2016) A comprehensive literature review on load frequency control strategies. Int J Sci, Eng Technol Res 5(11):3220–3229
46. Gupta AP, Archana (2015) A literature survey: load frequency control of two area power system using fuzzy controller. Int J Proresses Eng Manag Sci Hum 1(5):1–14
47. Singh G, Bala R (2011) Automatic generation and voltage control of interconnected thermal power system including load scheduling strategy. Int J Eng Adv Technol 1(2):1–7
48. Gaing Z-L (2004) A particle swarm optimization approach for optimum design of PID controller in AVR system. IEEE Trans Energy Convers 19(2):384–391
49. Dubey A, Bondriya P (2016) Literature survey on load frequency controller. Int Res J Eng Technol 3(5):1604–1611
50. John N, Ramesh K (2012) An overview of load frequency control strategies: literature survey. Int J Eng Res Technol 1(10):1–5
51. El Yakine Kouba N, Menaa M, Hasni M, Boudour M (2016) LFC enhancement concerning large wind power integration using new optimised PID controller and RFBs. IET Gener Transm Distrib 10(16):4065–4077
52. Falehi AD, Rostami M, Mehrjadi H (2011) Transient stability analysis of power system by coordinated PSS-AVR design based on PSO technique. Engineering 3(05):478–484
53. Manickam SR, Mustafa MTA (2014) Design of BAT inspired algorithm based dual mode gain scheduling of PI load frequency control controllers for interconnected multi-area multi-unit power systems. Aust J Basic Appl Sci 8(18):635–647
54. Sahu BK, Pati S, Panda S (2014) Hybrid differential evolution particle swarm optimisation optimised fuzzy proportional–integral derivative controller for automatic generation control of interconnected power system. IET Gener Transm Distrib 8(11):1789–1800
55. Ali ES (2014) Optimization of power system stabilizers using BAT search algorithm. Electr Power Energy Syst 61:683–690

56. Omar B, Saida IB (2014) Bat algorithm for optimal tuning of PID controller in an AVR system. International conference on control, engineering & information technology 2014, pp 158–170
57. Mohanty B, Panda S, Hota PK (2014) Differential evolution algorithm based automatic generation control for interconnected power systems with non-linearity. Alex Eng J 53(3):537–552
58. Sahu RK, Panda S, Padhan S (2014) Optimal gravitational search algorithm for automatic generation control of interconnected power systems. Ain Shams Eng J 5(3):721–733
59. Samuel GG, Christober Asir Rajan C (2015) Hybrid: particle swarm optimization–genetic algorithm and particle swarm optimization–shuffled frog leaping algorithm for long-term generator maintenance scheduling. Int J Electr Power Energy Syst 65:432–442
60. Sahu RK, Panda S, Padhan S (2015) A hybrid firefly algorithm and pattern search technique for automatic generation control of multi area power systems. Int J Electr Power Energy Syst 64:9–23
61. Sahu RK, Panda S, Chandra Sekhar GT (2015) A novel hybrid PSO-PS optimized fuzzy PI controller for AGC in multi area interconnected power systems. Int J Electr Power Energy Syst 64:880–893
62. Sahu BK et al (2015) Teaching–learning based optimization algorithm based fuzzy-PID controller for automatic generation control of multi-area power system. Appl Soft Comput 27:240–249
63. Chitara D et al (2015) Optimal tuning of multimachine power system stabilizer using cuckoo search algorithm. IFAC-PapersOnLine 48(30):143–148
64. Eke Ī, Taplamacıoğlu MC, Lee KY (2015) Robust tuning of power system stabilizer by using orthogonal learning artificial bee colony. IFAC-PapersOnLine 48(30):149–154
65. Özdemir MT et al (2015) Tuning of optimal classical and fractional order PID parameters forAutomatic generation control based on the bacterial swarm optimization. IFAC-PapersOnLine 48(30):501–506
66. Soni V et al (2016) Hybrid Grey wolf optimization-pattern search (hGWO-PS) optimized 2dof-Pid controllers for load frequency control (Lfc) in interconnected thermal power plants. ICTACT J Soft Comput 6(3):1244–1256
67. Gupta E, Saxena A (2016) Performance evaluation of antlion optimizer based regulator in automatic generation control of interconnected power system. J Eng 2016:1–14
68. Ahmed BS et al (2016) Optimum design of PIλDμ controller for an automatic voltage regulator system using combinatorial test design. PLoS One 11(11):1–20
69. Venkatesh M, Sudheer G (2017) OPTIMAL LOAD FREQUENCY REGULATION OF MICRO-GRID USING DRAGONFLY ALGORITHM. Int Res J Eng Technol 4(8):978–981
70. Pain S, Acharjee P (2017) Tuning of PID controller for realistic load frequency control system using chaotic exponential PSO algorithm. Int J Eng Technol 8(6):2712–2724
71. Kumar A, Suhag S (2017) Multiverse optimized fuzzy-PID controller with a derivative filter for load frequency control of multisource hydrothermal power system. Turk J Electr Eng Comput Sci 25(5):4187–4199
72. Sambariya DK, Fagna R (2017) A novel elephant herding optimization based PID controller design for Load frequency control in power system. In: 2017 International Conference on Computer, Communications and Electronics (Comptelix). IEEE, pp 595–600
73. Dey P, Bhattacharya A, Das P (2017) Tuning of power system stabilizer for small signal stability improvement of interconnected power system. Appl Comput Inform. https://doi.org/10.1016/j.aci.2017.12.004
74. Khezri R, Oshnoei A, Hagh MT, Muyeen SM (2018) Coordination of heat pumps, electric vehicles and AGC for efficient LFC in a smart hybrid PowerSystem via SCA-based optimized FOPID controllers. Energies 11:420–1–21
75. Mathur P, Rajpurohit VS, Srivastava RK (2018) Comparative analysis of PID tuning of AVR. Int J Futur Revol Comput Sci Commun Eng 4(1):53–56

Chapter 3
Self-Organization: A Perspective on Applications in the Internet of Things

Furqan Ahmed

3.1 Introduction

The concept of Internet of things (IoT) entails connecting everyday smart devices to the Internet, paving the way for a plethora of new applications and services, in both consumer and industrial domains. The devices include sensors, actuators, and controllers, connected using communication technologies such as Bluetooth, WiFi, ZigBee, LTE-NB, etc. Sensing of environment often results in generation of massive amounts of data, which can be analyzed and processed in the cloud. In general, the devices are heterogeneous in terms of communication, computation, intelligence, and storage capabilities. Thus, IoT can be envisioned as a highly complex and resource-constrained system of systems, which is powered by both real-time network intelligence and longtime analytics. Notable applications include smart environments, intelligent transportation and logistics, healthcare, and industrial automation. For a detailed overview of IoT applications in everyday life and industrial processes, see, for example, [1, 2], and references therein. The motivation for self-organization in IoT emanates from the applications and use-cases that require proximity-based wireless communications between neighboring devices, both with or without the involvement of fixed network infrastructure. Enabling the devices to communicate directly without dependence on centralized network infrastructure or full cloud connectivity is important from a latency, reliability, and resiliency perspective.

F. Ahmed (✉)
KTH Royal Institute of Technology, Stockholm, Sweden

Elisa, Helsinki, Finland
e-mail: furqan.ahmed@elisa.fi

X. Li, K.-C. Wong (eds.), *Natural Computing for Unsupervised Learning*, Unsupervised and Semi-Supervised Learning,
https://doi.org/10.1007/978-3-319-98566-4_3

The potential use-cases of self-organization in this direction are directly related to device-to-device (D2D) communication, in both standalone and network-controlled modes. For example, in vehicle-to-vehicle (V2V) communication scenarios, vehicles can exchange messages when in proximity, for enhanced safety and improved traffic flow [3]. Likewise, self-organizing algorithms are important tools for managing unmanned aerial vehicle traffic [4]. Emergency and public safety scenarios where the public infrastructure network is not available due to congestion or disasters constitute another important example. In such circumstances, self-organized functionalities can be employed to provide reliable communication and safety services to first responders as well as the end users [5]. Other use-cases include smart buildings and industrial automation. In smart buildings and offices, low-power short-range technologies such as Bluetooth can make use of self-organizing D2D functionalities to seamlessly connect a large number of heterogeneous nodes in a plug-and-play manner [6]. The use of self-organized D2D in industrial automation is motivated by latency and reliability constraints. In particular, industrial processes involving time-sensitive feedback control loops consisting of sensors controlling actuators require such solutions [7].

In the current 5G vision, a key enabler for IoT is machine-type communications (MTC) or machine-to-machine (M2M) paradigm, which consists of massive machine and mission-critical applications. Massive number of devices is a defining characteristic of IoT, with the total number expected to be around 50 billion by the year 2020. On the other hand, applications in industry, smart cities, public safety, and medicine often have mission-critical aspects. The mission-critical services may consist of performance targets in terms of latency, availability, and reliability. Due to the unprecedented potential of M2M paradigm in future networks especially IoT, it is important to understand its constituent scenarios and their respective challenges. We hold the view that a number of emerging M2M use-cases require self-organizing functionalities due to requirements such as scalability, rapid deployment, cost- and resource-efficiency. Apart from [5] and [8], the potential of self-organization for M2M in IoT networks has not been clearly addressed in the literature before. In order to bridge this gap, we present a unified overview of self-organization in M2M networks. To this end, we focus on use-cases involving devices communicating directly with each other, with little or no direct control from network infrastructure. Notable examples include mission-critical M2M use-cases such as vehicle-to-vehicle (V2V) communications and public safety networks (e.g., emergency services and critical infrastructure monitoring) and generic device-to-device (D2D) communications including low-power short-range devices forming capillary networks.

For V2V, the motivation for self-organization is highly dynamic topology and performance targets in terms of latency, whereas for public safety networks, rapid large-scale deployment and resiliency aspects of self-organization networking are important. Moreover, mission-critical scenarios also arise in industrial M2M scenarios involving real-time control loops with sensors controlling actuators over wireless links – a stark contrast to simplistic unidirectional communication scenarios such as sensors transmitting data to cloud. This motivates the enabling of distributed intelligence via self-organization mechanisms at network edge. Cloud-based centralized M2M approaches often assume that cloud connectivity is always available to all devices, latency is tolerable, cloud has sufficient computing power and storage to process all the data generated by devices, and continuous signaling from cloud is not straining the energy resources of devices communicating with it. Furthermore, there are also security and privacy concerns surrounding the concept of centralized clouds. In fact, such centralized solutions are not scalable and require new hierarchical and distributed approaches. Chiang et al. discuss in detail the motivation for moving away from centralized approaches toward distributed approaches on network edge [9]. In particular, mission-critical M2M challenges can be tackled by the fog computing paradigm, which consists of distributed computing platform comprising of local fog nodes with single hop connection to the end devices [10]. Moreover, these fog nodes can self-organize to form a mesh while maintaining connectivity to the cloud. A key advantage of fog is that based on data processing and computational capabilities, rapid decision-making can be delegated to geographically close smart devices acting as local controllers [8]. It is also more energy efficient compared to continuous communication with a centralized controller.

The M2M-enabled architecture for IoT can be envisioned as multiple layers, where low-level devices such as sensors are connected to sink nodes – devices with much higher computing and storage power (smartphones, gateways). Nevertheless, enabling smart devices to discover their neighboring peers and communicate with them constitutes an effective approach toward improving the resiliency of network. The benefits can be further enhanced by introducing mesh connectivity and multi-hop routing between devices. The cloud provides the connectivity for actions that occur over longer timescales, such as data analytics and network management. Therefore, self-organization is an effective paradigm in the fog and device layer, to handle diverse M2M use-cases at network edge, including V2V and smart homes. In the rest of this chapter, we present an overview of self-organization followed by a discussion on the use-cases and key benefits of self-organization for M2M in IoT. A technical overview of state of the art is also presented, which gives an insight into current self-organization approaches.

3.2 Self-Organizing Systems: An Overview

A key feature of self-organizing systems is the lack of central control for coordinating the actions of components in the system [11, 12]. The system is often distributed in the sense that individual components act as peers and each component interacts only with its neighbors. These interactions collectively drive the system to an ordered state. Thus, the system is inherently scalable and the underlying principles are well-suited to highly complex and large-scale dynamic systems. Moreover, the principles of self-organization have been successfully applied for solving complex problems in electrical engineering and computer science. Its popularity stems from the fact that it enables distributed solutions with limited communication, computation, and energy consumption at the system level. Self-organizing systems should involve local and limited coordination among system components. The main features of self-organizing systems that motivate the application of self-organization in communication networks are summarized next:

3.2.1 Scalability

An important property of self-organizing systems is scalability, and it is of utmost importance for large-scale network deployments such as sensor networks in IoT. It ensures that the system can be scaled up without an unbounded increase in complexity. Scalability emanates from the self-organization principle itself. Due to the local nature of interactions between the system components, addition of new components results in limited additional complexity. This is especially important for dynamic systems, which evolve over time in terms of size and number of components. A design principle based on local interactions yields scalable solutions for a wide range of problems. Local interaction between system components can be formulated in many different ways, depending on the nature of problem. For ideal self-organization, the local interaction should be entirely passive and based on mere observations of the neighboring components. Self-organization with entirely passive local interactions is mostly found in biological systems. However, when dealing with engineering of self-organization, e.g., in wireless networks, the local interactions may also be active and involve an exchange of data or cooperation in the form of message passing.

3.2.2 Adaptability

Self-organizing systems are adaptable and, therefore, can effectively respond to changes in the environment and adjust operation parameters accordingly, to maintain performance level. To this end, the system can reorganize in order to adapt to changes such as addition/removal of components in the system or loss in system performance resulting from other factors. In this regard, agility is also important as the system should be able to react to the changes in a timely manner to ensure that the performance remains unaffected. Thus, both adaptability and agility are important for guaranteeing rapid convergence to a desirable state.

3.2.3 *Emergence*

Another important property of self-organizing systems is that they exhibit the phenomenon of convergence to an equilibrium, i.e., the local interactions among system components lead to an emergence of desirable properties at the global level. All the system components contribute toward the desirable global behavior but are not capable of achieving it individually. The collective intelligence of system components leads to an emergence of global behavior that transcends the capabilities of individual components.

3.2.4 *Resilience*

The adaptable and decentralized nature of self-organizing systems makes them particularly resilient against component failures. This is due to the fact that system can adapt to changes in the environment and has the capability to reconfigure itself in the case of failure. The decentralized architecture is intrinsically robust and leads to properties such as stability and graceful degradation. In order to enhance the robustness, self-organizing systems are often designed with self-healing mechanisms to tackle failures in a systematic manner, thereby making the system more resilient.

3.3 Self-Organizing Wireless Networks

3.3.1 *Self-Organization in Contemporary Cellular Networks*

In contemporary cellular networks such as Long-Term Evolution (LTE) and LTE Advanced (LTE-A), the need for self-organizing mechanisms arises due to network densification and heterogeneity, which makes the configuration, optimization, and maintenance of the network difficult, thereby leading to higher capital expenditure (CAPEX) and operational expenditure (OPEX) [13]. Apart from CAPEX and OPEX reduction, network automation using self-organized networking (SON) can address challenges of ensuring seamless coverage and capacity, scalability, efficient radio resource management, energy efficiency, fault tolerance, and quick recovery. The primary objective of SON in cellular networks is to minimize human intervention in network design and operation. In addition, network automation through approaches are capable of adapting to varying network topology, channel conditions, and quality of service requirements of the users. SON leads to better network performance, higher scalability, and improved robustness.

A number of SON use-cases have been considered by third-generation partnership project (3GPP) in the standardization of LTE and LTE-A. These use-cases have been identified in different research projects such as Self- Optimization and Self-Configuration in Wireless Networks (SOCRATES) [14] and standardization activities of 3GPP [15]. Moreover, a number of important use-cases have been proposed by Next Generation of Mobile Communications (NGMN) [16].

The SON use-cases related to network deployment and operation are usually classified into three main categories: self-configuration, self-optimization, and self-healing. Self-configuration of network nodes involves automated download and installation of software followed by configuration of basic network parameters. This plug and play functionality is especially important for heterogeneous networks (HetNets) comprising of random and uncoordinated deployments of low-power nodes, where traditional network planning approaches are not feasible. On the other hand, self-optimization entails measurement of network conditions and key performance indicators, which are subsequently used to optimize network parameters. Examples include coverage and capacity optimization (CCO), intercell interference coordination, and optimization of mobility and load balancing parameters. Self-healing functionalities aimed at minimizing performance degradations are triggered when network detects performance deterioration such as outages due to node failures.

Some important SON use-cases considered by 3GPP from Release 8 onwards are as follows [13]:

- Automatic neighbor relations.
- Automated physical-cell ID assignment.
- Mobility robustness optimization (MRO).
- Random access channel optimization.
- Mobility load balancing (MLB).
- Energy savings (ES).
- Intercell interference coordination.

In 3GPP standardization activities of cellular networks, both centralized and distributed SON has been considered. Therefore, SON algorithms may be fully distributed that do not involve message passing, distributed approaches with message passing, or even centralized mechanisms of network operation. Centralized SON means that all SON algorithms and functions are located in operations and management system, at a high hierarchical level. In distributed SON, all algorithms and functions are located at a relatively lower level, i.e., at evolved Node B level. Architecture comprising of combination of a set of SON functions located at different hierarchical levels is called hybrid SON [23].

Moreover, the time scales involved depend on the nature of the problem. Time scales for CCO and ES use-cases may be in a range of few minutes to hours, whereas for MRO, channel-aware scheduling, and interference mitigation, it would be milliseconds to seconds. A classification of SON based on time scales is given in [17]. In SCNs, different SON parameters may change at very different time scales ranging from milliseconds/seconds to days/months [18].

A number of resource allocation problems pertinent to IoT can be addressed through discrete and continuous network optimization methods, where the aim is to achieve a network-level objective under various constraints. This approach has been considered for cellular SON problems such as capacity enhancement [19], coverage optimization [20], dynamic frequency allocation [21], spectrum assignment [22], resource allocation for self-healing [23], and interference mitigation [24]. In most cases, distributed optimization methods are employed to design self-organizing algorithms for SCNs. The SON problems such as physical-cell ID assignment and primary component carrier selection can be addressed by graph coloring approaches. On the other hand, for problems involving continuous variables, network utility maximization (NUM) is an effective approach. The NUM framework provides an impetus for a systematic design of self-organized resource allocation in communication networks, and it can enable both distributed and fully distributed SON algorithms. Distributed algorithms are cooperative in nature and involve an exchange of dedicated messages between the cells. Messages may simply communicate the transmission parameters of a given node or some more sophisticated parameter related to a NUM problem, e.g., prices defined to compute the loss in utility of a given node, resulting from a change in another nodes transmission parameters. The prices can be exchanged through dedicated message passing between the network nodes or over a backhaul link in the case of cellular networks. Fully distributed algorithms can be obtained by simply setting the price terms equal to zero. An alternative approach toward fully distributed SON is motivated by ad hoc networking applications, where each node broadcasts a network-level price over the air. The nodes in the neighborhood can listen and adjust their parameters accordingly.

3.3.2 Self-Organization for Enabling Internet of Things

In future 5G networks, apart from throughput-centric small-cell deployments, proximity-based services involving heterogeneous D2D communication such as V2V and public safety would require enhanced self-organizing capabilities. Moreover, self-organizing D2D could play an important role in meeting latency and resiliency constraints in massive machine communications and mission-critical communications. Related self-organization problems include resource allocation, coverage and capacity maximization, interference management, and maintenance of quality of service and connectivity in the network. Another emerging area that motivates the self-organizing capabilities in 5G is nomadic deployments where there is uncertainty regarding their spatial and temporal availability. The network nodes are installed dynamically when there is a need for improved network performance. These nodes constitute an effective mechanism of providing good connectivity in emergency scenarios and other sporadic situations such as outdoor events. Most importantly, self-organization has always been an important component in the design of sensor and actor networks. This trend is likely to come to the forefront in the 5G era, paving the way for self-organizing IoT. In the context of 5G, self-organizing features under consideration for IoT include self-configuration, self-optimization, self-healing, self-protection, and self-processing of data [25, 26]. Apart from network resilience, self-organizing approaches can enable efficient allocation of resources (e.g., energy, spectrum, computational power, storage) in D2D systems based on local and contextual information, thereby playing a pivotal role in improving overall performance. Self-organizing mechanisms can be employed for intelligent resource allocation, spectrum usage, and power consumption of these devices [27]. Thus, self-organization is becoming increasingly pertinent and has the potential of becoming an important enabler of M2M networks in the 5G vision. In what follows, we discuss three M2M use-cases that require self-organization.

3.4 Self-Organizing M2M for IoT: Use Cases

3.4.1 Home M2M

Home networking is an important area related to IoT, witnessing a transition from human-to-machine to M2M paradigm. This is due to an increase in household devices and appliances embedded with sensors and different low-power wireless networking technologies. A main feature of home networks is device heterogeneity – as the devices range from very low-power battery-operated devices such as fire alarms and temperature sensors to high-power appliances connected to main electricity outlets. Likewise, the devices vary in terms of computation resources (e.g., smartphones, laptops, and personal computers). The devices connect in a self-organizing manner, without any human intervention, using standard low-power wireless technologies such as Bluetooth, ZigBee, WiFi, etc. For instance, Zhang et al. discusses a scenario comprising of multiple such self-organizing subnetworks, including body area networks, entertainment network, and convenience networks [28]. These self-organizing subnetworks may be based on different technologies and connect to home gateway via an intelligent sub-gateway. Thus, self-organized mesh networking of smart sensors (e.g., lighting, motion, temperature, humidity, and security) can enable highly efficient and integrated solutions.

Cognitive M2M-based home area networks for efficient energy management and demand response management is discussed in [29]. It comprises of a cognitive gateway, smart meters, sensors, actuators, and other intelligent devices interacting in a self-organizing manner to reduce the peak load. A home energy management systems for smart grid is proposed in [30]. It makes use of clustering among the nodes to find optimal traffic concentration in the network. In this context, the design of topology formation and reliable mesh networking protocols for home networks comprising of heterogeneous devices is also an important issue worth investigating.

3.4.2 V2V Communications

Vehicle-to-vehicle (V2V) and vehicle-to-infrastructure communication constitutes an important self-organizing M2M use-case. It involves vehicles exchanging short and frequent messages with nearby vehicles or roadside units, to enhance safety and ensure a smooth journey. Vehicle-to-vehicle control is a typical self-organization problem, as the vehicles have to act in an autonomous manner, to coordinate their actions based on local information and latency constraints. For instance, the vehicles connect to each other when they are in the communication range and exchange information such as speed, direction, and location. High data rate applications include content sharing and see-through vehicle where a vehicle streams its dash cam video to the vehicle behind it to enhance safety. Gerla et al. discusses self-organization in the context of vehicular fogs [31]. The authors give a detailed account of the evolution from intelligent vehicle grid to autonomous vehicles and vehicular fogs.

Moreover, the vehicles can form a multi-hop ad hoc network, where a vehicle acts as a router forwarding messages to distant cars and road infrastructure. This can enhance not only the safety but also efficiency of traffic flow. The motivation for cognitive M2M in intelligent transportation system is presented in [29]. In such system, M2M is used to connect vehicles, traffic lights, as well as road side units. Intelligent sensors and data processing devices can be used to make the roads cognitive and interconnected leading to significant improvement in traffic efficiency and safety. The use of M2M communications for enhanced driving is discussed in [32], in which authors propose a future transportation scenario comprising of vehicles communicating information such position, velocity, maps, and driver state. Cars receiving such information can respond by taking actions such as messages, honking, or communicating warning messages to other cars and pedestrians to avoid accidents and enhance safety.

3.4.3 Public and Infrastructure Safety

Self-organization is well-known approach toward increasing network resilience against natural and man-made hazards. Therefore, it can be considered as an important tool for public safety M2M scenarios involving outages, attacks, and disasters. Such networks must have very short deployment times along with self-healing features to maintain connectivity without human intervention, in the case of network infrastructure and node failures. Athreya et al. gives an overview of the concept of self-organization in IoT and focuses on network resiliency aspects relevant to maintaining connectivity in emergency situations and disasters [5]. The components identified for the design of self-organizing networks include neighbor discovery, medium access control, local connectivity and path establishment,

Table 3.1 Self-organizing M2M use-cases for IoT

M2M use-cases	Benefits of self-organization	Applications	Reference
Home M2M	Low cost, high resilience	Smart grid and demand response management, entertainment and security systems, smart living	[28–30]
V2V communications	Highly dynamic topology, low latency	Accident avoidance, autonomous driving, content sharing.	[29, 31, 32]
Public and infrastructure safety	Rapid deployment, high resilience, low latency	Disaster management, emergency services, surveillance, monitoring	[5, 33–35]

service recovery management, and energy management. An overview of enabling technologies for IoT and an algorithm approach for IoT-based post-disaster response and recovery are proposed in [33]. It consists of ad hoc network formation using D2D-based communication between devices. Likewise, in [34], a peer-to-peer cloud network service for IoT-based disaster information is developed. Here, the idea is to combine IoT-enabled M2M network with P2P cloud service for rapid response in case of disaster and to provide the results as social services. Further examples of such systems developed using swarm intelligence principles can be found in [35]. The self-organizing M2M SON use-cases for IoT discussed above are summarized in Table 3.1, where main motivation of self-organization, applications, and representative existing work are also highlighted. It is worth noting that self-organizing M2M networks have diverse applications that share a common ground in that devices communicate locally in an automated manner without centralized controller.

3.5 Toward Self-Organizing Intelligent IoT

In self-organizing systems, local interactions among system components lead to desired behavior at global level. It is an effective tool to handle the challenges posed by complex networks such as scalability, latency, rapid and low-cost deployment, and resilience. These factors impact the design of network management, optimization, and control. In principle, regarding overall network architecture, there are three main approaches, namely, centralized, distributed, and hybrid. The centralized approaches enable more efficient and intelligent decisions but obviously lack scalability due to a centralized controller and therefore are not suitable for many M2M applications. In particular, long-ranged communication with centralized controller may lead to higher latency and rapid consumption of energy. Distributed approaches on the other hand suffer from communication-control trade-off, meaning that the more active nodes communicate and exchange information, the better is the network control and performance. Insufficient internode communication in distributed networks may also lead to problems in convergence.

The objective behind the application of self-organization to M2M is to enable distributed intelligence and cognitive capabilities in network, in an autonomous manner. The decisions made by the self-organizing devices are local, in the sense that devices usually have a limited view of network. However, these local interactions have an influence on the global behavior of the system. This leads to an interesting question: How to enable network-level intelligent behavior in M2M systems by designing local actions of individual devices? Two concepts are important in this regard – self-organizing systems and cognitive systems. In cognitive systems, the focus is on cognitive cycle comprising of understanding, learning, and reasoning. On the other hand, in self-organizing networks intelligence primarily emerges from the local interactions of network nodes, which may or may not involve cognitive cycle. In general, M2M in 5G systems will benefit from a confluence of self-organization and cognitive networking [36].

A few studies have discussed potential applications of self-organization for M2M in IoT systems. Ding et al. proposed self-organizing scheme based on an endocrine regulation mechanism, for service discovery [37]. The hormonal exchange acts as a mechanism via which information is exchanged between the nodes. Thus, nodes cooperate by exchanging information locally. An overview of bio-inspired algorithmic approaches for self-organization in contemporary cellular networks and M2M is given in [35]. Wu et al. discuss cognitive IoT, which relies on cognitive cycle, to enable intelligence [36]. It also possesses a context-aware component, which further improves the learning process. Context-aware computing is of paramount importance in this regard, especially in the case where nodes have sensing capabilities resulting in massive data generation. An in-depth survey of context-aware computing for IoT systems is given in [38]. Learning approaches for self-organization are discussed in [39] and [40]. These are based on reinforcement learning and sequential learning. In addition, the authors also introduce cognitive hierarchy theory for handling the device heterogeneity. The benefits of self-organization can be further exploited by its application to fog computing, as discussed in [9]. Moreover, an in-depth survey of economic and pricing models is presented [26]. A related network utility maximization approach for resource allocation in devices sharing unlicensed spectrum is discussed in [41].

A summary of self-organization approaches toward enabling M2M use-cases in IoT discussed above is given in Table 3.2. It can be seen that a wide range of methods including artificial intelligence and machine learning can be applied to enable network intelligence in self-organizing M2M networks.

Table 3.2 Different approaches toward intelligence in self-organizing M2M

Network intelligence mechanism	Problem	Use-case	Reference
Bio-inspired approaches, artificial intelligence (artificial endocrine system)	Resource allocation, service discovery, MAC and network layer issues	Use-cases involving dynamic topologies, heterogeneous networks	[35, 37]
Machine learning (supervised, unsupervised, reinforcement)	Resource allocation, wireless sensor networks	Use-cases involving dynamic and distributed optimization	[39, 42, 43]
Cognitive systems (learning, thinking, sensing)	Resource allocation, automatic network operation, service provisioning	Intelligent transportation, smart grid	[29, 36]
Economic and pricing models	Resource allocation, sensing data collection	Generic crowd sensing and M2M, spectrum sharing devices in unlicensed bands	[26, 41]

3.6 Conclusions

In this chapter, the concept of self-organization and its relevance to IoT is discussed. Starting with an overview of self-organization, we reflect on the motivation and applications of self-organization in contemporary and future cellular networks. A number of important SON use-cases in contemporary LTE/LTE-A cellular network are discussed. Next, we discuss the motivation for self-organized networking in 5G and IoT networks. Three use-cases for self-organization in M2M networks for IoT are discussed. These include home M2M networks where devices connect to enable smart living, vehicle-to-vehicle networks for improved safety and content sharing, and public infrastructure and safety networks that enable rapid response to disasters. All these use-cases required proximity-based direct communications between devices in a self-organizing manner. Finally, we give an overview of important current approaches for enabling intelligent self-organizing solutions for IoT.

References

1. Atzori L, Iera A, Morabito G (2010) The internet of things: a survey. Comput Netw 54(15):2787–2805
2. Xu LD, He W, Li S (2014) Internet of things in industries: a survey. IEEE Trans Ind Inform 10(4):2233–2243
3. Abd-Elrahman E, Said AM, Toukabri T, Afifi H, Marot M (2015) A hybrid model to extend vehicular intercommunication v2v through d2d architecture. In: International conference on computing. Networking and Communications (ICNC), Anaheim

4. Virgh C, Nagy M, Gershenson C, Vsrhelyi G (2016) Self-organized uav traffic in realistic environments. In: International conference on intelligent robots and systems (IROS), Deajeon
5. Athreya AP, Tague P (2013) Network self-organization in the internet of things. In: International workshop of internet-of-things networking and control (IoT-NC), New Orleans
6. Sachs J, Beijar N, Elmdahl P, Melen J, Militano F, Salmela P (2014) Capillary networks–a smart way to get things connected. Ericsson Rev 8(2014):1–8
7. Samad T (2016) Control systems and the internet of things. IEEE Control Syst 36(1):13–16
8. Prazeres C, Serrano M (2016) SOFT-IoT: self-organizing FOG of things. In: 30th international conference on advanced information networking and applications workshops (WAINA), Crans-Montana
9. Chiang M, Zhang T (2016) Fog and IoT: an overview of research opportunities. IEEE Internet Things J 3(6):854–864
10. Bonomi F, Milito R, Zhu J, Addepalli S (2012) Fog computing and its role in the internet of things. In: Workshop on mobile cloud computing, New York
11. Haken H (2006) Information and self-organization: a macroscopic approach to complex systems. Springer-Verlag, New York
12. Camazine S (2003) Self-organization in biological systems. Princeton University Press, Princeton
13. Hämäläinen S, Sanneck H, Sartori C (2012) LTE self-organising networks (SON): network management automation for operational efficiency. Wiley, West Sussex, UK
14. SOCRATES (2008) D2.1 Use cases for self-organizing networks
15. 3GPP (2008) TS 36.902 v0.0.1 Evolved universal terrestrial radio access; self-configuring and self-optimizing network use cases and solutions (Release 8)
16. Lehser F (2008) Next generation mobile networks—use cases related to self organizing networks, overall description, NGMN Alliance
17. Fehske AJ, Viering I, Voigt J, Sartori C, Redana S, Fettweis GP (2014) Small-cell self-organizing wireless networks. Proc IEEE 102(3):334–350
18. Hossain E, Le LB, Niyato D (2013) Radio resource management in multi-tier cellular wireless networks. Wiley, Hobonken, New Jersey
19. Akbarzadeh S, Combes R, Altman Z (2012) Network capacity enhancement of ofdma system using self-organized femtocell off-load. In: Wireless communications and networking conference (WCNC)
20. Ho LTW, Ashraf I, Claussen H (2009) Evolving femtocell coverage optimization algorithms using genetic programming. In: 20th international symposium on personal, indoor and mobile radio com- munications
21. Chowdhury MZ, Jang YM, Haas ZJ (2010) Interference mitigation using dynamic frequency re-use for dense femtocell network architectures. In: Second international conference on ubiquitous and future networks
22. Wu Y, Jiang H, Zhang D (2012) A novel coordinated spectrum assignment scheme for densely deployed enterprise lte femtocells. In: 75th vehicular technology conference (VTC spring), Yokohama
23. Lee K, Lee H, Cho DH (2011) Collaborative resource allocation for self-healing in self-organizing networks. In: International conference on communications (ICC)
24. Lima CHMD, Bennis M, Latva-aho M (2011) Coordination mechanisms for stand-alone femtocells in self-organizing deployments. In: IEEE global telecommunications conference, Kathmandu
25. Palattella MR, Dohler M, Grieco A, Rizzo G, Torsner J, Engel T, Ladid L (2016) Internet of things in the 5G era: enablers, architecture, and business models. IEEE J Sel Areas Commun 34(3):510–527
26. Luong NC, Hoang DT, Wang P, Niyato D, Kim DI, Han Z (2016) Data collection and wireless communication in internet of things (iot) using economic analysis and pricing models: a survey. IEEE Commun Surv Tutorials 18(4):2546–2590
27. Bello O, Zeadally S (2016) Intelligent device-to-device communication in the internet of things. IEEE Syst J 10(3):1172–1182

28. Zhang Y, Yu R, Xie S, Yao W, Xiao Y, Guizani M (2011) Home M2M networks: architectures, standards, and QoS improvement. IEEE Commun Mag 49(4):44–52
29. Zhang Y, Yu R, Nekovee M, Liu Y, Xie S, Gjessing S (2012) Cognitive machine-to-machine communications: visions and potentials for the smart grid. IEEE Netw 26(3):6–13
30. Niyato D, Xiao L, Wang P (2011) Machine-to-machine communications for home energy management system in smart grid. Commun Mag 49:53–59
31. Gerla M, Lee EK, Pau G, Lee U (2014) Internet of vehicles: from intelligent grid to autonomous cars and vehicular clouds. In: World forum on internet of things (WF-IoT)
32. Healey J, Wang C-C, Dopfer A, Yu C-C (2012) M2M gossip: why might we want cars to talk about us?, ACM, New York, NY, USA
33. Kamruzzaman M, Sarkar NI, Gutierrez J, Ray SK (2017) A study of IoT-based post-disaster management. In: International conference on information networking, Da Nang
34. Chung K, Park RC (2016) P2P cloud network services for IoT based disaster situations information. Peer-to-Peer Network Appl 9(3):566–577
35. Zhang Z, Long K, Wang J, Dressler F (2014) On swarm intelligence inspired self-organized networking: its bionic mechanisms, designing principles and optimization approaches. IEEE Commun Surv Tutorials 16(1):513–537
36. Wu Q, Ding G, Xu Y, Feng S, Du Z, Wang J, Long K (2014) Cognitive internet of things: a new paradigm beyond connection. IEEE Internet Things J 1(2):129–143
37. Ding Y, Jin Y, Ren L, Hao K (2013) An intelligent self-organization scheme for the internet of things. IEEE Comput Intell Mag 8(3):41–53
38. Perera C, Zaslavsky A, Christen P, Georgakopoulos D (2014) Context aware computing for the internet of things: a survey. IEEE Commun Surv Tutorials 16(1):414–454
39. Park T, Abuzainab N, Saad W (2016) Learning how to communicate in the internet of things: finite resources and heterogeneity. IEEE Access 4:7063–7073
40. Park T, Saad W (2016) Learning with finite memory for machine type communication. In: Information science and systems
41. Ahmed F, Tirkkonen O (2009) Local optimum based power allocation approach for spectrum sharing in unlicensed bands. In: International workshop on self-organizing systems. Springer, Berlin/Heidelberg
42. Alsheikh MA, Lin S, Niyato D, Tan H-P (2014) Machine learning in wireless sensor networks: algorithms, strategies, and applications. IEEE Commun Surv Tutorials 16(4):1996–2018
43. Aliu O, Imran A, Imran M, Evans B (2013) A survey of self organisation in future cellular networks. IEEE Commun Surv Tutorials 15(1):336–361

Part II
Advances in Unsupervised Learning

Chapter 4
Applications of Unsupervised Techniques for Clustering of Audio Data

Surendra Shetty and Sarika Hegde

4.1 Introduction

Audio data mining is an important area of research in the field of computing technology. These techniques can be used for automatically analyzing and searching/indexing the content of an audio data. The growing amount of digital contents in the world of computers has increased the challenges in managing and retrieving the audio data [8]. The advancements in technology in this area of audio data mining have made the scientists to develop more beneficial applications helpful to society [6]. Content-based music retrieval is one of the significant applications of audio data mining, which makes use of the content such as melody, instrument, rhythms, etc. for efficient management of music repository [17]. The study of audio data mining includes the study of various fields like statistics, pattern recognition, neural networks, artificial intelligence, signal processing, machine learning, and data mining. Machine learning is a part of artificial intelligence, and to be intelligent, a system that is in a changing environment should have the ability to learn. If the system can learn and adapt to such changes, the system designer need not foresee and provide solution for all possible situations [5]. Classification and clustering are the major techniques used under audio data mining.

In this chapter, we discuss the various clustering techniques like K-means, fuzzy C-means, hierarchical algorithms, and Gaussian mixture models (GMM). These techniques are used for finding the groupings or clustering of the different types of audio data using different feature sets, presuming no prior knowledge of natural groupings or classes in the dataset. The dataset for the experiments includes the audio signals comprising Carnatic classical music and animal voice. For the animal

S. Shetty (✉) · S. Hegde
NMAM Institute of Technology, Nitte, Karnataka, India
e-mail: hsshetty@nitte.edu.in

X. Li, K.-C. Wong (eds.), *Natural Computing for Unsupervised Learning*, Unsupervised and Semi-Supervised Learning,
https://doi.org/10.1007/978-3-319-98566-4_4

voice dataset, we have collected the various animals' voice listed as *alligator*, *bear*, *blackbird*, *cardinal*, *cat*, *cow*, *donkey*, *elephant*, *frog*, *monkey*, *peacock*, *snake*, *tiger*, *and zebra* (Source such as http://www.ilovewavs.com/Effects/Animals/Animals.htm). In music data, we have considered two types of datasets as listed below:

- Collection of music data obtained from different Indian instruments like *flute*, *harmonium*, *mandolin*, *nagaswaram*, *santoor*, *saxophone*, *sitar*, *shehnai*, *veena*, and *violin* (monophonic audio clips collection from CDs).
- Classical songs rendered by different *South-Indian* singers: *Balamuralikrishna*, *Nithyashree*, *Sudha Ranganathan*, *Unnikrishnan*, *Vasanthakumari*, and *Yesudas* (monophonic audio clips collection from CDs).

4.2 Structure of Audio Data

Electronically, sounds are represented as either analog or digital signals. In analog form, sound is a continuous waveform. Digital signals are in discrete form and popularly used in storage and computational processing purpose. Digital signals are created from the analog signal using sampling and quantization technique by retrieving the air pressure value from analog signal at regularly spaced time interval. The air pressure value is also specified as amplitude or sample in digital signal processing (DSP) field. The word sample is the popular term used in DSP for indicating that it is a sampled amplitude value taken from a continuous signal [8]. Sampling rate is defined as the number of times the amplitude of the signal is measured per second. The higher the sampling rate, the more accurately the sampled signal will represent the original signal. For speech, the sampling rate may vary like 8000, 11,000, 22,000 samples per second as it is generally considered in signal processing literature. The most common sampling rate used for music is 44,100 samples per second. For example, music data of 3 s duration can be digitally stored with $44{,}100 \times 3 = 1{,}32{,}300$ samples. If we have 30,000 sample, then the duration of music (with sampling rate of 44,100) is $T = 30{,}000/44100 = 0.75$ s. This calculation can be used to compute the duration of audio if we know the number of samples [17].

4.2.1 *Mathematics of Audio Signals*

A continuous signal is represented by its amplitude $x(t)$ at time t. When a continuous signal is converted into digital signal, it is represented by $x[n]$, representing the amplitude value of nth sample observation. The conversion is represented as $x[n] = x(nT_s)$, T_s representing a sampling interval with a frequency of $\frac{1}{T_s}$ Hz. The vector $x[n]$ can be termed as audio vector which is a sequence of sample observations. Thus, $X = (x_1, x_2, \ldots\ldots, x_N)$, where N is the total number of digitized sampled observations, represents an audio vector of N sample values. For any

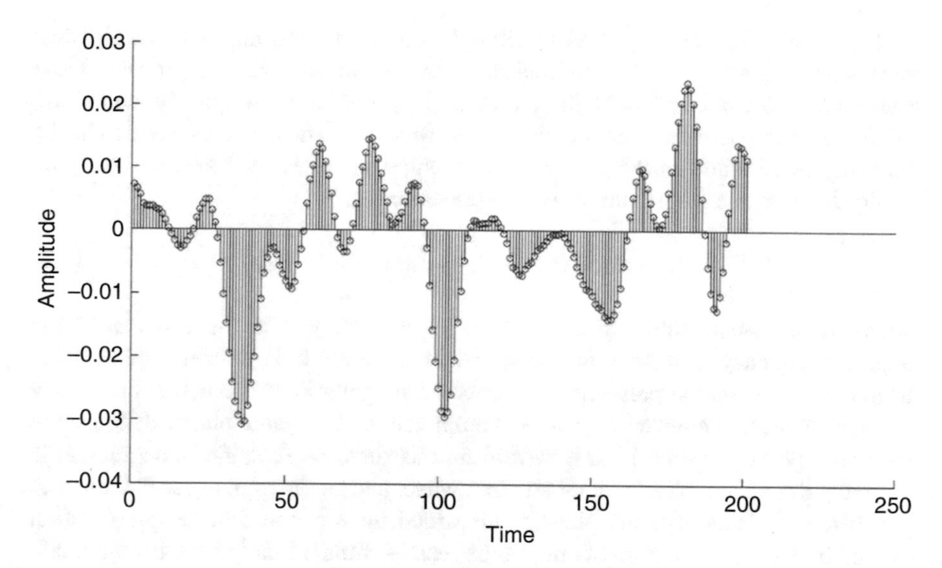

Fig. 4.1 Plot of amplitude over time for 250 samples of an audio signal

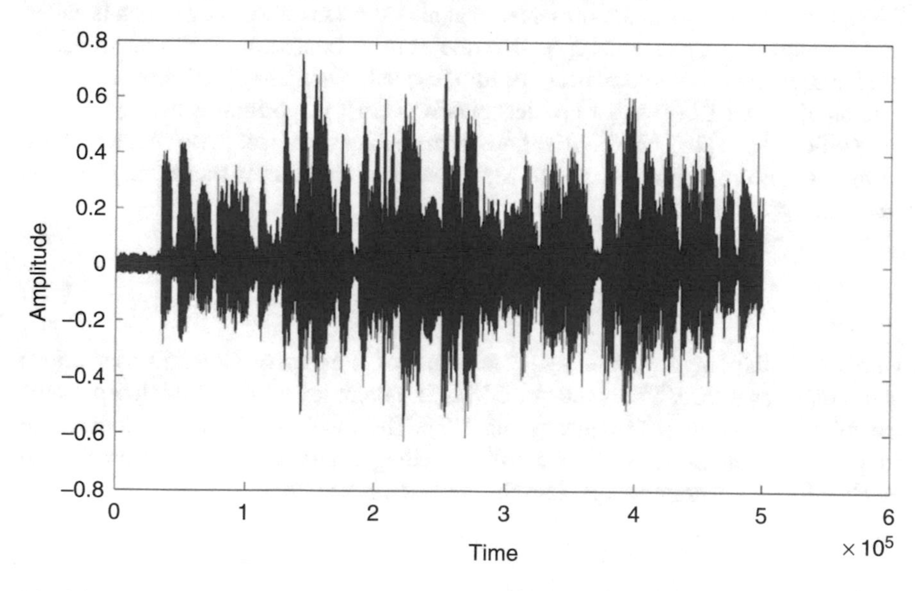

Fig. 4.2 Plot of amplitude over time for 5 lakhs sample observations of audio signal

audio signal, if we draw the samples over discrete time on 2-d plane, we could see a sinusoidal shape (a sine/cosine wave) rising and falling. The following graph (Fig. 4.1) shows one such example for 250 samples of an audio signal [11]. Similarly, the graph in Fig. 4.2 shows the plot of 5 lakhs samples with sampling rate 44.1 KHz, and the duration of the audio clip is 11.36 s.

Digital signal processing (DSP) techniques can be directly applied on audio data represented as sequence of amplitudes, or the audio data can be processed and represented as combination of three factors, i.e., amplitude, frequency, and phase which are the parameters of the sine wave function. Thus, audio signals can be logically represented as the composition of sine/cosine wave. Mathematically an audio signal with a single sine wave is represented as:

$$x(t) = A \sin(\omega t + \Phi) = A \sin\left(2\pi f t + \Phi\right) \tag{4.1}$$

where A represents the amplitude, f is the frequency, t is the time, ω is the angular frequency, and Φ is the phase angle. Amplitude is defined with respect to normal atmospheric pressure also called as magnitude of the wave. Frequency is defined as the number of cycles completed in 1 s, and phase denotes the particular point in the cycle of a waveform, measured as an angle in degrees. It is normally not an audible characteristic of a wave, and so the human ear is known as "phase-deaf." Therefore, the phase is discarded most of the time. Representation of audio data as variation of amplitude versus time is called as time-domain signal denoted as $X[n]$.Similarly, audio data can be decomposed into a number of frequency components and represented magnitude versus frequency which is called as frequency-domain signal, $X[f]$. Discrete Fourier transform (DFT) technique is used to convert the time-domain signal into frequency-domain signal. Generally, one version of the DFT called fast Fourier transform (FFT) algorithm is used as it is the most efficient one [16]. The Fourier transform enables to extract these characteristics from a waveform $x[n]$ and maps to $X(f)$. The discrete Fourier transform (DFT) is calculated on sampled signal as:

$$X\left[k\right] = \sum_{n=0}^{N} x\left[n\right] e^{-ikn2\pi} / N \tag{4.2}$$

where N is the length of the signal in terms of number of sample observations and k represents the k^{th}frequency bin. $X[k]$ is a complex number which represents the magnitude of the k^{th}frequency bin [13]. The DFT is plotted by plotting the frequencies along x-axis and the amplitude along y-axis (it is common practice to take the log of the amplitude); phase information is usually discarded.

4.2.2 Preprocessing of Audio Data

In preprocessing step, audio data is simplified so that the features can be computed directly on this. We describe techniques like framing and windowing which are significant in audio preprocessing.

4.2.2.1 Framing

Framing is a process of breaking the set of samples of an entire audio file into smaller chunks called as frame. The framing is done with overlapping where each frame of the audio data is made to contain part of the previous frame and part of the next frame instead of straightforward segmentation. This ensures that audio features occurring at a discontinuity are at least considered whole in subsequent, overlapped frame [8]. The audio vector X is divided into sequence of frames $F = (f_1, f_2, \ldots, f_Q)$, where Q indicates the number of frames in F. Each value f_i indicates a frame which is a sub-array of sample observation values. In the process of framing, truncation problem may occur which is overcome by using the process called windowing.

4.2.2.2 Windowing

A window function is a mathematical function that is zero-valued outside of some chosen interval. When another function or a signal (data) is multiplied by a window function, the product is also zero-valued outside the interval. All that is left is the part where they overlap which is the "view through the window." This process is called as windowing, and the size of the window is same as size of the frame. There are many defined windows like Hamming, Hanning, rectangular, triangular, etc. The coefficients of L-point symmetric Hanning window are computed from the following equation [8].

$$w\left[k+1\right] = 0.5\left(1 - \cos\left(2\pi \frac{k}{L-1}\right)\right), \qquad k = 0, \ldots\ldots, L-1 \qquad (4.3)$$

The weights can be stored in vector $W = w(k)$, $k = 0, \ldots, L - 1$. In the same way, the Hamming window coefficients are computed as follows:

$$w\left[k+1\right] = 0.54 - 0.46\cos\left(2\pi \frac{k}{L-1}\right), \qquad k = 0, \ldots\ldots, L-1 \qquad (4.4)$$

Each frame f_i is multiplied with the window W to avoid the truncation problem in the signal.

4.2.3 Feature Extraction

Feature extraction is the process of transformation of raw data from one form to another. It is the compact representation of the original data and should uniquely represent the patterns in the original data. The procedure used to derive the features from the data is called as feature extraction technique. Depending on the type of information required from audio, the feature extraction technique varies. For speech signal, DSP techniques are used to retrieve the features. Some of the speech features are directly computed from time-domain signal, or some are retrieved from frequency-domain signal [13].

Feature extraction technique is applied for each of the windowed frame f_i, containing L samples of the audio file, and the frame gets converted into either a single value or a vector containing sequence of values/coefficients. Figure 4.3 shows the flowchart illustrating the method of converting the audio file into feature vector FV. If the feature has more than one attributes or coefficient, it is represented as feature vector $\mathrm{FV} = \{\ \mathrm{fv}_1, \mathrm{fv}_2 \cdots, \mathrm{fv}_d\}$ where 'd' is the total number of features or total number of coefficients/attributes of single feature. Sometimes the features in the original data sets have the necessary information, but it is not in the form suitable for the data mining algorithms. In such situations, one or more new features are constructed out of the original feature FV.

If the number of features is very large, then further processing may become difficult. To overcome this problem, dimensionality reduction can be applied where it reduces the dimensionality of the data set by creating new attributes that are a combination of the old attributes. We can also apply the step called as feature subset selection, where the reduction of dimensionality is done by selecting a subset of attributes which is a subset of the original feature set [4]. The set of features have been mainly categorized into two types, low-level features and high-level content description features. Computation of low-level features is already described above. The high-level features are based on the knowledge/domain expertise about the piece of audio computed using low-level features [3]. We will discuss some of the popular feature extraction techniques, namely, mel-frequency cepstral coefficients (MFCC), linear predictive coefficients (LPC), short-time energy, zero-crossing rate, and spectral centroid.

4.2.3.1 Mel Frequency Cepstral Coefficients (MFCC)

MFCC is a feature extracted by applying more than one Fourier transform to the original signal in sequence. The term cepstrum (an anagram of spectrum) derives from this repeated use of the fast Fourier transform (FFT). The MFCC feature seeks to emulate human perception through the following general procedure, as shown in Fig. 4.4 [6, 12]. The first step is preprocessing which consists of framing and windowing of the signal. For each of the frame f_i, DFT coefficients are calculated by applying FFT algorithm. After applying the DFT, the resulting power spectrum is transformed to mel-frequency scale. This is done by using a filter bank consisting of triangular filters, spaced uniformly on the mel scale. An approximate conversion between a frequency value in Hertz (f) and *mel* is given by [9]:

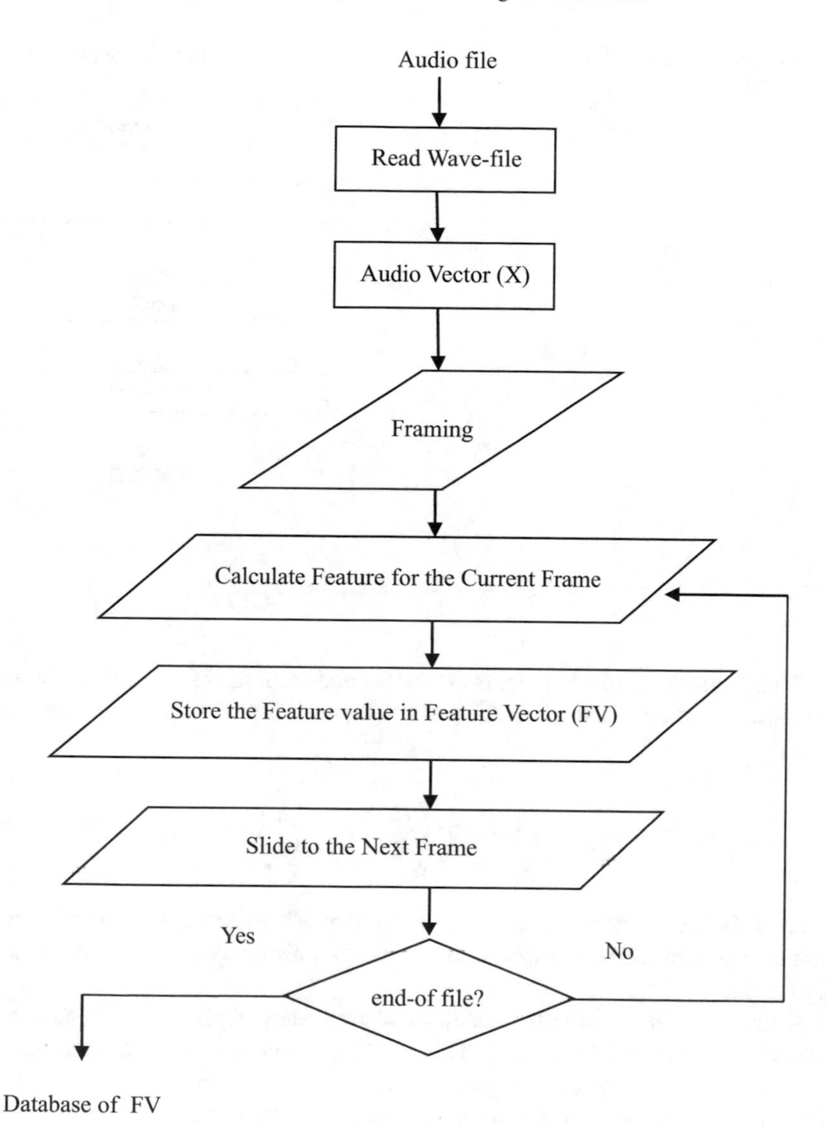

Fig. 4.3 Process of feature extraction

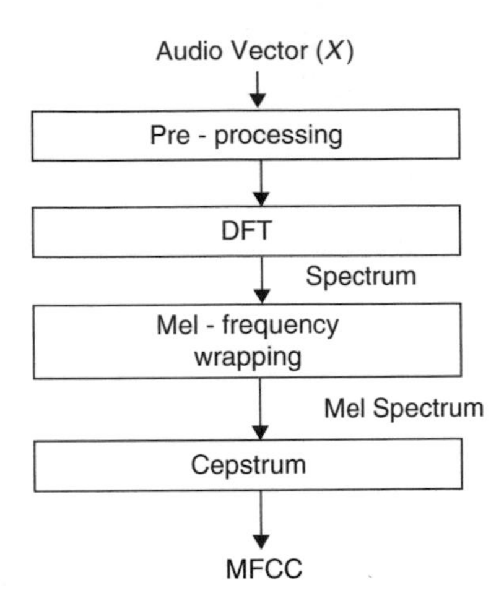

Fig. 4.4 MFCC computation for a signal

$$\text{mel}(f) = 2595\log_{10}\left(1 + \frac{f}{700}\right) \tag{4.5}$$

Finally, the cepstral coefficients are calculated from the Mel-spectrum by taking the discrete cosine transform (DCT) of the logarithm of the mel spectrum. This is given by

$$C_i = \sum\nolimits_{k=1}^{K} (\log S_k) \, . \cos\left(\frac{i\pi}{K}\left(k - \frac{1}{2}\right)\right) i = 1, 2, \ldots, K \tag{4.6}$$

where $C_i$is the ith MFCC, S_k is the output of kth filter bank channel (i.e., the weighted sum of the power spectrum bins on that channel), and K is the number of filterbanks.

A more detailed description of the mel-frequency cepstral coefficients can be found in Rabiner and Jaung [10]. This final representation is an approximation of the compressed and equalized signal produced by the mechanism of human hearing. Given the correlation, it makes an attractive feature for context-aware research and has been found to be very discriminatory [10]. The more MFCCs, the more precise the approximation of the signal's spectrum, and it also means more variability of the data [2]. Aucouturier and Pachet [2] have used *eight* coefficients. Out of the 13 coefficients computed using Auditory Toolbox, we have considered the first 6 coefficients for the experiments [12]. The following sections Ex1 and Ex2 gives the Matlab code for framing, windowing, and computation of MFCC feature for few frames.

Ex 1: Matlab Code for Framing and Windowing of an Audio Data

```
%Framing and windowing of audio file
clear all
pathd='../Datasets/InstrumentAudio/test/flute1.wav';

%Read the wavefile
[x fs]= wavread(pathd);

%Length of audio vector
len=length(x(:,1));

%Intialize the frame size and overlapping size (50% of frame_size)
frame_size=1000;
inc=frame_size/2;

%Number of frames that we may get
nof=((len-1)/inc)+1;

%Intialize start and end position for 1st frame
st=1;
en=st+frame_size-1;

%Read each frame by incrementing the start and end position
i=1;
while en<len
   f(i,:)=x([st:en],1)';
   st=st+inc;
   en=st+frame_size-1;
   i=i+1;
end
%Remaining samples are added to f by adding
f(i,[1:frame_size])=zeros(frame_size,1)';
f(i,[1:len-st+1])=x([st:len],1)';

%Apply hanning window for one frame
win=hann(frame_size);
   w=(f(1,:))'.*win;
  end
```

Ex 2: Compute MFCC for each Frame

```
%Call melcepst function for first frame
fprintf('MFCC value for first frame');
cep=melcepst(f(1,:), fs)

%Call melcepst function for first frame
fprintf('MFCC value for second frame');
cep=melcepst(f(2,:), fs)

%Call melcepst function for fifteenth frame
fprintf('MFCC value for fifteenth frame');
cep=melcepst(f(50,:), fs)
```

Output

MFCC value for first frame

cep = 6.7898 -3.2180 0.6366 -4.1934 -0.2703 -3.8028 -0.6507 -1.9286 0.9199 -0.1660 1.8003 0.8421

MFCC value for second frame

cep = 6.8387 -3.8423 -0.3045 -4.2899 -0.6234 -4.1339 -0.5999 -2.3309 0.9273 -0.0104 1.9160 0.4668

MFCC value for fifteenth frame

cep = 10.5932 -3.8120 1.2710 -3.6195 -1.4647 -1.8923 -3.3520 -0.2316 -0.0819 1.7484 0.6864 1.1201

Consider the animal voice data with six categories of animals, namely, *alligator*, *cat*, *bear*, *elephant*, *monkey*, and *tiger*. MFCC is extracted from each frame of size 50 ms for all the audio files. We randomly select 25 observations (frames) from each of the class (animal type) and plotted the first three MFCC coefficients (fv_{i1}, fv_{i2}, fv_{i3}) in 3D-space as shown in the Fig. 4.5. It can be observed that observations from same animal type are grouped together. This visualization helps us to understand that the MFCC may be used as one of the features for identifying the type of the animal from the given audio.

Similar experiments are conducted for identifying the instruments and the singers in audio clips. The instrument data of *ten* categories and singer data of six categories are considered. The plots are generated with first three MFCCs of randomly selected 25 observations belonging to each of the instrument type shown in Fig. 4.6. We observe from the plot that the observations belonging to same type of instruments (indicated with same symbols) are forming a group as they are nearer to each other compared to those of other types.

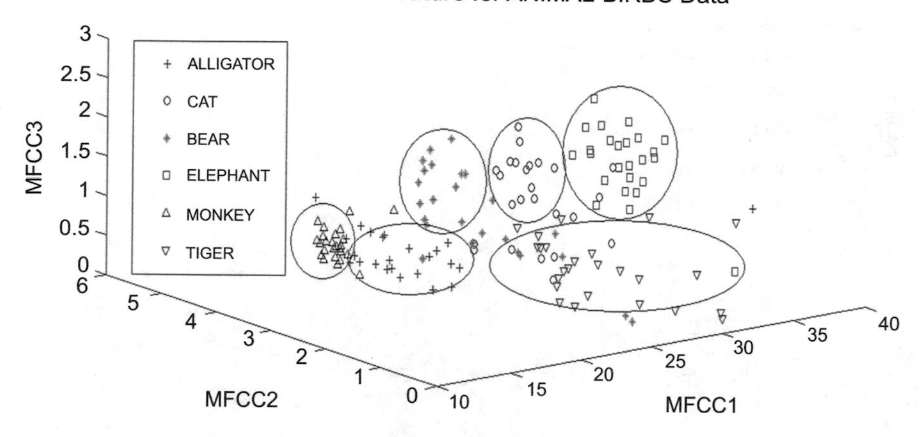

Fig. 4.5 Graph for MFCC feature for animal voice data

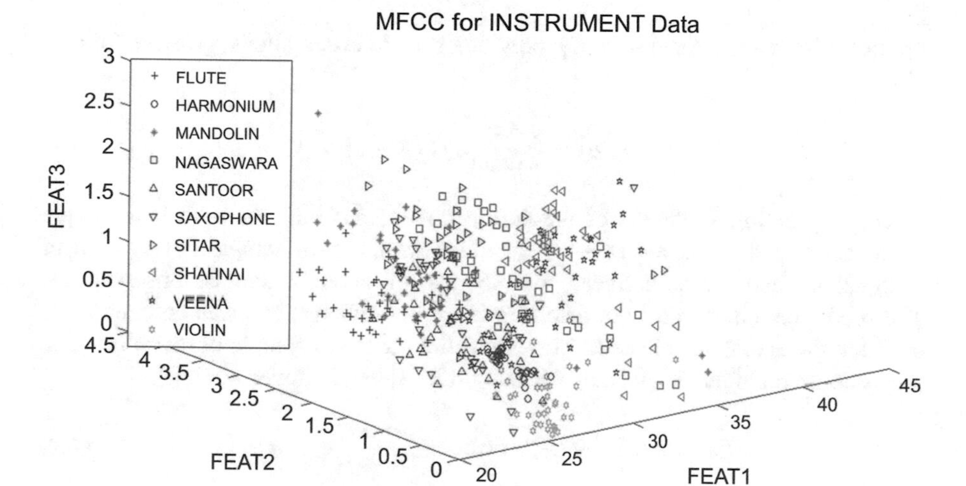

Fig. 4.6 MFCC feature for instrument data

Similar plot is generated for singer audio data as shown in Fig. 4.7. We observe that there is some similarity between the observations of the same type even though the distinction is not very strong.

4.2.3.2 Linear Predictive Coefficients (LPC)

Linear predictive analysis is one of the most commonly used speech analysis tool. It has also been used in filter design, spectral analysis, and system identification. The basic idea behind linear prediction is that the next signal sample observation is

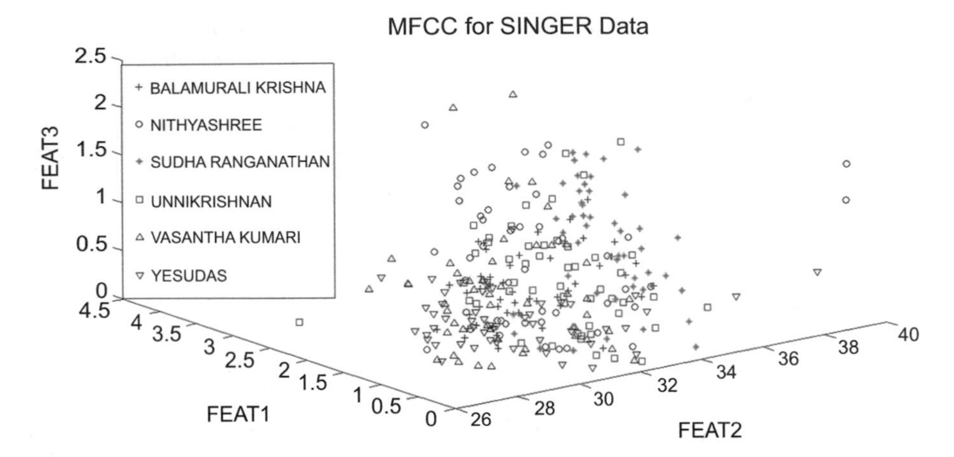

Fig. 4.7 MFCC for singer data

predicted from a weighted sum of p previous sample observations, given as follows [9]:

$$\widehat{s}(\boldsymbol{n}) = \sum\nolimits_{i=1}^{P} a_i s(\boldsymbol{n} - \boldsymbol{i}) \tag{4.7}$$

where the set $\{a_i\}$ is the set of prediction coefficients and $s(\boldsymbol{n} - \boldsymbol{i})$ is a sample observation at time instant $(\boldsymbol{n} - \boldsymbol{i})$. In other words, each sample observation of a signal is modeled as a linear combination of previous sample observations. The prediction coefficients are determined by minimizing the mean squared error between the actual sample observation and the predicted sample observation. The prediction error signal is given by McLoughlin [8] and Peltonen et al. [9]

$$\boldsymbol{e}(\boldsymbol{n}) = \boldsymbol{s}(\boldsymbol{n}) - \widehat{s}(\boldsymbol{n}) = \boldsymbol{s}(\boldsymbol{n}) - \sum\nolimits_{i=1}^{p} a_i s(\boldsymbol{n} - \boldsymbol{i}) \tag{4.8}$$

and the squared error over all the $\boldsymbol{n}$ sample observations:

$$\boldsymbol{E} = \sum\nolimits_{n} [\boldsymbol{e}(\boldsymbol{n})]^2 = \sum\nolimits_{n} \left[s(\boldsymbol{n}) - \sum\nolimits_{i=1}^{p} a_i s(\boldsymbol{n} - \boldsymbol{i}) \right]^2 \tag{4.9}$$

The technique described above computes '$\boldsymbol{p}$' LPC coefficients for a given frame of size $\boldsymbol{n}$.

We have considered the first *6* LPC coefficients out of *13* coefficients computed using Auditory Toolbox. For each frame f_i,the sequence of *six* LPC coefficients is given as $\mathbf{LP} = (\mathbf{lp_1}, \mathbf{lp_2}, \mathbf{lp_3}, \mathbf{lp_4}, \mathbf{lp_5}, \mathbf{lp_6})$. LP is used as the feature vector (**FV**) for discriminating different classes of audio signal. Figure 4.8 shows the result of plotting the LPC coefficients for randomly selected 25 observations from each of the animal type. These results are plotted in 3D space for the first three coefficients

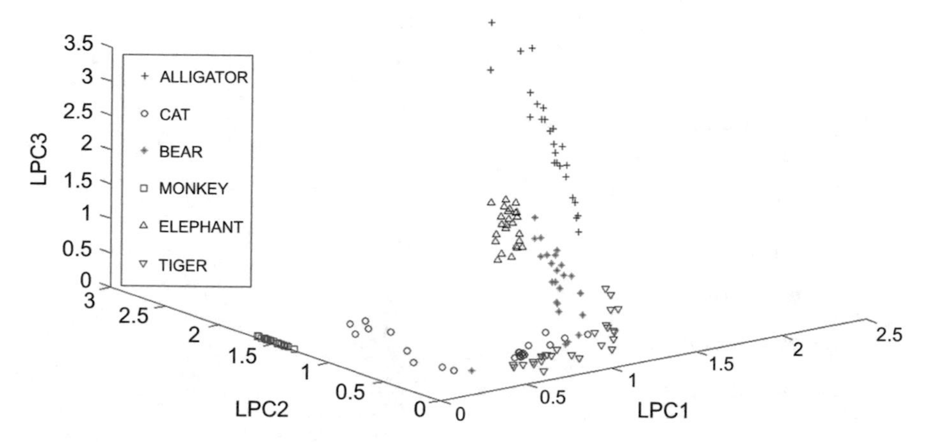

Fig. 4.8 LPC feature for animal data

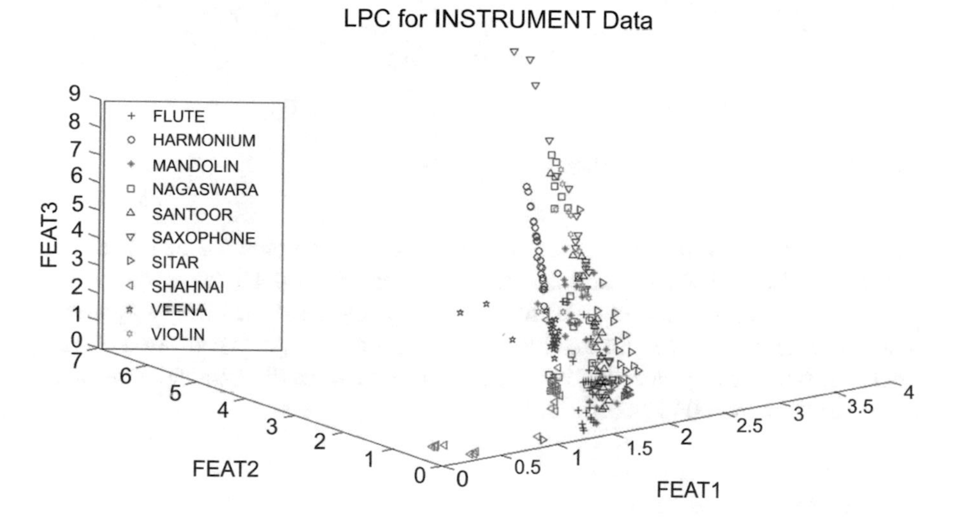

Fig. 4.9 LPC feature for instrument data

$(\mathbf{lp_1}, \mathbf{lp_2}, \mathbf{lp_3})$, and we observe that LPC also helps in grouping the objects of same type together. It can be observed that the LPC coefficients are equally good compared to MFCC coefficients.

Similar experiments were conducted for *ten* types of instruments as shown in Fig. 4.9. The observations belonging to same instrument type (indicated with same symbols) are grouped together.

It can be observed that similar types of instruments like *saxophone* (inverted triangle) and *nagaswaram* (circle) groups are nearer to each other. This shows the possibility of using the LPC coefficients as one of the promising feature for

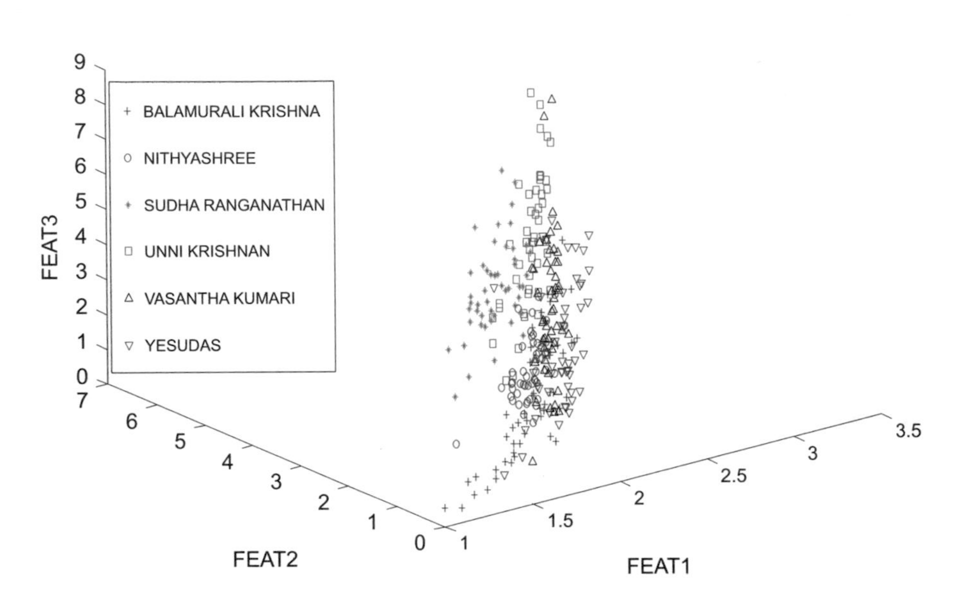

Fig. 4.10 LPC feature for singer data

discriminating the different instrument classes. The discrimination of instruments through MFCC is slightly clearer than LPC features. Figure 4.10 shows the result of plotting the first *three* LPC coefficients that are randomly selected from singer data. We can observe from the plot that the discrimination of singer types based on LPC coefficients is not very clear, but it is better compared to the discrimination of singer types based on the MFCC coefficients (Fig. 4.7).

4.2.3.3 Short-Time Energy Function

This feature is calculated for time-domain signal. For a given frame, the short-time energy ($\boldsymbol{E}$) can be directly computed as the root mean square of audio vector, [8] given by

$$\boldsymbol{E} = \sqrt{\frac{\sum_{i=0}^{L-1} x_i{}^2}{L}} \tag{4.10}$$

Figure 4.11 shows the result of plotting the short-time energy values for the randomly selected sample of 30 observations from each class of the animal voice

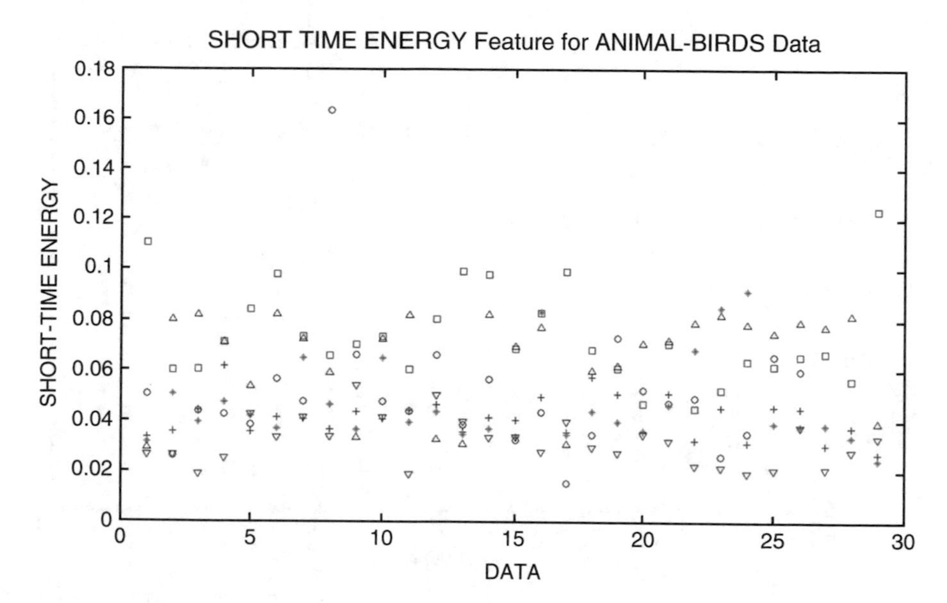

Fig. 4.11 Short-time energy

data. As we can observe, the data belonging to the same animal voice type do not show much similarity for the energy feature.

4.2.3.4 Zero-Crossing Rate

Zero-crossing rate (ZCR) is given as the number of times the time-domain signal crosses the x-axis for a given frame [8]. The ZCR value for a given frame is calculated as [9].

$$\mathbf{ZCR} = \frac{1}{2N}\left(\sum_{i=1}^{N-1}\left|\text{sgn}\left(\boldsymbol{x}\left(i\right)\right) - \text{sgn}\left(\boldsymbol{x}\left(i-1\right)\right)\right|\right) \tag{4.11}$$

where $\boldsymbol{x(i)}$ represents the $\boldsymbol{i^{th}}$ sample observation and $\boldsymbol{x(i-1)}$ represents the $\boldsymbol{(i-1)^{th}}$ sample observation with $\boldsymbol{N}$ samples in the frame. The sign function is defined as follows:

$$\text{sgn}\left(\boldsymbol{x}\right) = \begin{cases} \mathbf{+1}, & \boldsymbol{if x > 0} \\ \mathbf{0}, & \boldsymbol{if x = 0} \\ \mathbf{-1}, & \boldsymbol{if x < 0} \end{cases} \tag{4.12}$$

The above method generates the feature $\mathbf{fv}_i$, which represents the value of the ZCR for one frame. Figure 4.12 shows the values of ZCR plotted for 30 observations

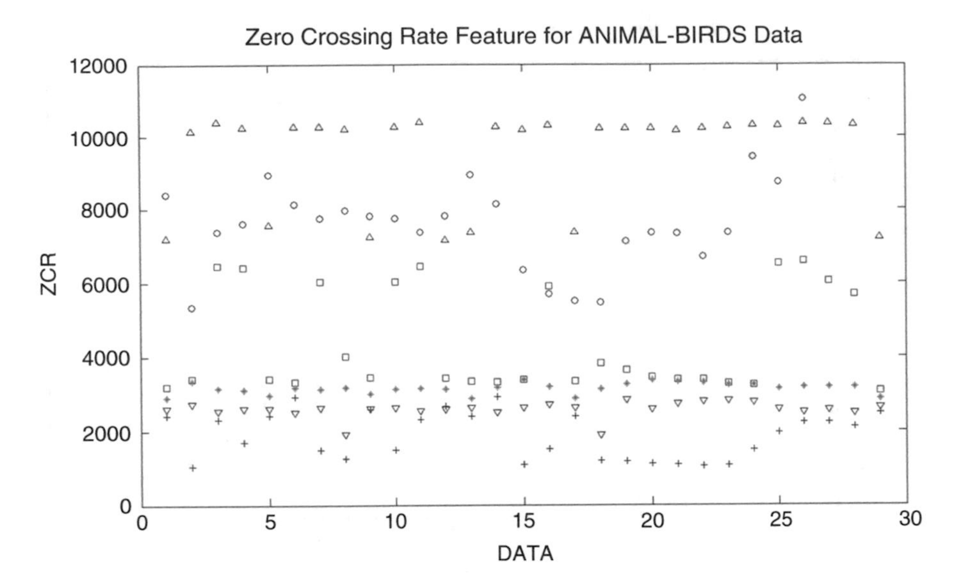

Fig. 4.12 Zero-crossing rate

randomly selected from each type of animal voice data. We can observe from the graph that the data belonging to same animal voice type are almost aligned in the same line. This indicates the similarity in the values of the ZCR for same type of animal voice.

4.2.3.5 Spectral Centroid

It represents the midpoint of the spectral power distribution which is related to the "brightness" of a sound. The higher value of spectral centroid indicates the higher brightness of the sound. To calculate the spectral centroid of the frame containing $\boldsymbol{L}$ sample observations, the frame is converted from time-domain signal to frequency-domain signal using the DFT technique. The spectral centroid for a frequency-domain signal (power spectrum) of a frame with $\boldsymbol{M}$ as the highest frequency is computed as [9]

$$\mathbf{SC} = \frac{\sum_{n=0}^{M} n.\left|X(n)\right|^2}{\sum_{n=0}^{M}\left|X(n)\right|^2} \tag{4.13}$$

Each value $\boldsymbol{X}(\boldsymbol{n})$ represents the magnitude of the nth frequency bin of the signal in frame. M is the index of the highest frequency bin. This method generates the feature $\mathbf{fv}_i$, which represents the value of the spectral centroid for one frame. Figure 4.13 shows the values of spectral centroid plotted for 30 observations randomly selected

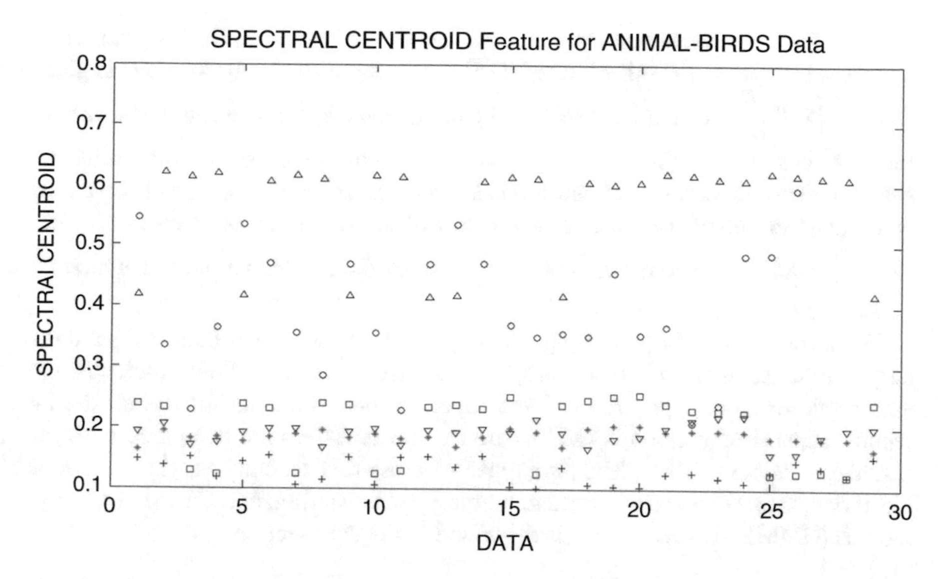

Fig. 4.13 Spectral centroid

from each type of animal voice data. It can be observed that the spectral centroid values of same type of animal voice are almost aligned in the same line. Hence, spectral centroid feature can be used as a distinguishing feature for discriminating animal voices.

4.3 Clustering of Audio Data

Clustering is similar to classification in which data is grouped, but unlike classification, the groups are not predefined. The objective of cluster analysis is to group the objects according to similarities among them. It is the process of assigning set of observations into subsets (called clusters) so that observations in the same cluster are similar in some way. Clustering techniques are the *unsupervised* methods; they do not use prior class identifiers. The main potential of clustering is to detect the underlying structure in data, which can be used not only for classification and pattern recognition but for model reduction and optimization. The clustering problem can be defined as follows:

Definition: Let X be our dataset, that is, $X = \{x_1, x_2, \ldots, x_N\}$. We define *k-clustering* of X as the partition of X into k sets (clusters) C_1, C_2, …, C_k, so that the following *three* conditions are met:

1. $C_i \neq \varnothing, i = 1, \ldots, k$
2. $\cup_{i=1}^{k} C_i = X$
3. $C_i \cap C_j = \varnothing, i \neq j, i, j = 1, \ldots, k$

In addition, the vectors contained in cluster $\boldsymbol{C_i}$ are "more similar" to each other and "less similar" to the other clusters [15]. The dataset for N observations is given as $\boldsymbol{D} = \left\{\mathbf{FV}^i\right\}_{i=1}^{N}$, constructed from audio data, where $\mathbf{FV}^i$ represents a feature vector for $\boldsymbol{i^{th}}$ observation. The clustering algorithm identifies the similarity/dissimilarity between the observations and constructs k clusters, where each of the observations is assigned to one of the clusters. The result of the clustering is given as a vector $\boldsymbol{O} = \left\{\mathbf{IDX}^i\right\}_{i=1}^{N}$, where $\mathbf{IDX}^i \epsilon \{\mathbf{1} : \boldsymbol{k}\}$ indicates the cluster number to which $\boldsymbol{i^{th}}$ observation belongs.

We have applied the clustering techniques to animal voice data, singer data, and instrument data. This study has been carried out to find how efficiently the audio data of same type are clustered together. For evaluation of the clustering results, supervised method is used where the cluster labels are compared with the actual known class labels. The *four* types of clustering techniques, viz., K-means clustering, fuzzy C-means clustering, hierarchical clustering, and Gaussian mixture models (GMM), are used, which is discussed in the next section.

4.3.1 Hierarchical Clustering

This algorithm produces a hierarchy of clusters. Initially, each observation is considered as an individual cluster, so we have totally N clusters in the first step. Then at each step, the two clusters with the shortest distance are merged into one, creating a new cluster, and correspondingly the distance of the new cluster with the rest is calculated. This process is repeated until we have $\boldsymbol{k}$ clusters. The steps in hierarchical algorithm are as follows:

1. Calculate the Euclidian distance between each pair of observations. We get totally $\boldsymbol{N} * (\boldsymbol{N} - \mathbf{1})$ unique pairs and the corresponding distances.
2. Group the objects into a binary, hierarchical cluster tree: In this step, we link pairs of objects that are in close proximity. The distance information generated in step 1 is used to determine the proximity of objects to each other. As objects are paired into binary clusters, the newly formed clusters are grouped into larger clusters until a hierarchical tree is formed.
3. Cut the hierarchical tree into clusters: In this step, *cluster* function is used which constructs a maximum of K clusters using distance as a criterion. Each node's height in the tree represents the distance between the two sub-nodes merged at that node. The function finds the smallest height at which a "horizontal cut" through the tree will leave K clusters [7].

4.3.2 K-means Clustering

K-means clustering is partitioning method of clustering. Here we specify $\boldsymbol{K}$, a user-specified parameter, namely, the number of clusters desired. We first chose $\boldsymbol{K}$ observations from dataset randomly and assign it as the $\boldsymbol{K}$ initial centroids. For each observation, we compute Euclidian distance between the observation and all the centroids. Then, each observation is assigned to the centroid with the smallest distance. The following algorithm shows the steps for K-means clustering [1]:

1. Select $\boldsymbol{K}$ points as initial centroids.
2. Repeat:
   (a) Form $\boldsymbol{K}$ clusters by assigning each point to its closest centroid.
   (b) Recompute the centroid of each cluster.
3. Until centroids do not change.

4.3.3 Fuzzy C-means Clustering

This is actually a fuzzy version of K-means algorithm, which is called fuzzy C-means. It is a data clustering technique wherein each data point belongs to a cluster to some degree that is specified by a membership weight, $\boldsymbol{w_{ij}} \in [\boldsymbol{0}, \boldsymbol{1}], \boldsymbol{j} = \boldsymbol{1}, \ldots, \boldsymbol{k}$, which indicates the value of membership for $\boldsymbol{i^{th}}$ observation in $\boldsymbol{j^{th}}$ cluster. For an observation $\mathbf{FV}^i$

$$\sum\nolimits_{j=1}^{k} w_{ij} = 1 \tag{4.14}$$

It starts with an initial guess for the cluster centers, which are intended to mark the mean location of each cluster. The initial guess for these cluster centers is most likely incorrect. Additionally, it assigns every observation a membership weight for each cluster. By iteratively updating the cluster centers and the membership weight for each observation, it iteratively moves the cluster centers to the right location within a data set. This iteration is based on minimizing an objective function that represents the distance from any given data point to a cluster center weighted by that data point's membership weight [14].

The steps for fuzzy C-means are as follows:

1. Select an initial fuzzy pseudo-partition, i.e., assign values to all the $\boldsymbol{w_{ij}}$randomly.
2. Repeat:
   (a) Compute the centroid of each cluster using the fuzzy pseudo-partition.
   (b) Recompute the fuzzy pseudo-partition, i.e., the $\boldsymbol{w_{ij}}$.
3. Until centroids do not change.

In the first step of initialization, weights are chosen randomly subject to the constraint that the weights associated with any object must sum to 1. For a cluster, C_j, the corresponding centroid, c_j, is computed as follows:

$$c_j = \sum_{i=1}^{N} w_{ij} x_i \Big/ \sum_{i=1}^{N} w_{ij} \quad (4.15)$$

In the last step, recomputing the fuzzy pseudo-partition, weights are updated. The weight w_{ij} associated with i^{th} observation and j^{th} cluster is recomputed as follows:

$$w_{ij} = \frac{1/\mathbf{dist}(x_i, c_j)^2}{\sum_{q=1}^{m} \left(1/\mathbf{dist}(x_i, c_j)^2\right)} \quad (4.16)$$

4.3.4 Gaussian Mixture Model Clustering

In this method, it is assumed that the data has been generated from a statistical model and then the data is described by finding the statistical model that best fits the data. At a higher level, this process involves deciding on a statistical model for the data and estimating the parameters for that distribution. Mixture models view the dataset as a set of observations from a mixture of different probability distributions [15]. Assume that there are k distributions and N observations:

$$X = \{x_i, i = 1, \ldots, N\}.$$

Let the jth distribution have parameters θ_j, and let Θ be the set of all parameters, i.e., $\Theta \equiv [\theta_1, \theta_2 \cdots, , \theta_k]$. Then, $\mathbf{prob}(x_i | \theta_j)$ is the probability of ith observation coming from the jth distribution. The probability that j^{th} distribution is chosen to generate an observation is given by the weight w_j $1 \le j \le k$, where these weights (probabilities) are subject to the constraint $\sum_{j=1}^{k} w_j = 1$. Then, the probability of an observation x_i is given by

$$\mathbf{prob}\left(x_i | \Theta\right) = \sum_{j=1}^{k} w_j \mathbf{prob}\left(x_i | \theta_j\right) \quad (4.17)$$

Since the observations are generated in an independent manner, then the probability of the entire dataset is just the product of the probabilities of each individual x_i:

$$\mathbf{prob}\left(X | \Theta\right) = \prod_{i=1}^{N} \mathbf{prob}\left(x_i | \Theta\right) \quad (4.18)$$

The probability density function for Gaussian distribution at a point $\boldsymbol{x}_i$ is

$$\mathbf{prob}\left(\boldsymbol{x}_i|\boldsymbol{\Theta}\right) = \frac{\mathbf{1}}{\sigma\sqrt{2\pi}}\boldsymbol{e}^{-\frac{\left(x_i-\mu\right)^2}{2\sigma^2}}, \quad -\infty < \boldsymbol{x}_i < \infty, \boldsymbol{\sigma} > \mathbf{0} \tag{4.19}$$

The parameters of the Gaussian distribution are given by $\boldsymbol{\Theta} = (\boldsymbol{\mu}, \boldsymbol{\sigma})$, where $\boldsymbol{\mu}$ is the mean of the distribution and $\boldsymbol{\sigma}$ is the standard deviation. We can estimate the parameters of these distributions from the data and describe these distributions (clusters).

4.4 Cluster Evaluations

We discuss the supervised method of cluster evaluation where the usual procedure is to measure the degree of correspondence between the cluster labels and the class labels. There are two kinds of approaches; the first set of techniques, called as classification-oriented, use measures from classification such as entropy and purity. These measures evaluate the extent to which a cluster contains objects of a single class. To compute entropy, the class distribution of the data for each cluster is calculated first, i.e., for cluster $\boldsymbol{j}$, we compute $\boldsymbol{p}_{ij}$, the probability that a member of cluster $\boldsymbol{i}$ belongs to class $\boldsymbol{j}$ as $\boldsymbol{p}_{ij} = \boldsymbol{m}_{ij}/\boldsymbol{m}_i$, where $\boldsymbol{m}_i$ is the number of observations in cluster $\boldsymbol{i}$ and $\boldsymbol{m}_{ij}$ is the number of observations of class $\boldsymbol{j}$ in cluster $\boldsymbol{i}$. Using these class distributions, the purity of each cluster is calculated as

$$\boldsymbol{p}_i = \mathbf{max}_j \boldsymbol{p}_{ij} \tag{4.20}$$

The overall purity is calculated as

$$\mathbf{purity} = \sum\nolimits_{i=1}^{\mathrm{K}} \frac{\boldsymbol{m}_i}{\boldsymbol{N}} \boldsymbol{p}_i \tag{4.21}$$

The entropy of each cluster $\boldsymbol{i}$ is given by

$$\boldsymbol{e}_i = -\sum\nolimits_{j=1}^{K} \boldsymbol{p}_{ij} \log_2 \boldsymbol{p}_{ij} \tag{4.22}$$

The total entropy for a set of clusters is calculated as the sum of the entropies of each cluster weighted by size of each cluster:

$$\boldsymbol{e} = \sum\nolimits_{i=1}^{K} \frac{\boldsymbol{m}_i}{\mathrm{N}} \boldsymbol{e}_i \tag{4.23}$$

The second set of methods is similarity-oriented, such as Jaccard measure and Rand statistic. These approaches measure the extent to which two observations that are in the same class are in the same cluster and vice versa. The two matrices, i.e., ideal cluster similarity matrix and ideal class similarity matrix, are compared. The ideal cluster similarity matrix is defined as below:

$$\mathbf{cluster}_{\text{matrix}}\left[i, j\right] = \left\{ \begin{array}{ll} 1 \text{ if } i^{\text{th}} \text{ and } j^{\text{th}} \text{observations are in the same cluster} \\ 0 \qquad\qquad\qquad\qquad\qquad\qquad \text{otherwise} \end{array} \right\}$$

Ideal class similarity matrix is:

$$\text{class}_{\text{matrix}}\left[i, j\right] = \left\{ \begin{array}{ll} 1 \text{ if } i^{\text{th}} \text{ and } j^{\text{th}} \text{observations are in the same class} \\ 0 \qquad\qquad\qquad\qquad\qquad\qquad \text{otherwise} \end{array} \right\}$$

To compute the binary similarity between the two matrices, the following four quantities are measured:

f_{00} = number of pairs of observations having a different class and a different cluster
f_{01} = number of pairs of observations having a different class and the same cluster
f_{10} = number of pairs of observations having the same class and a different cluster
f_{11} = number of pairs of observations having the same class and the same cluster.

After computing the above parameters, the simple matching coefficients known as the Rand statistic and the Jaccard coefficient, two of the most frequently used cluster validity measures, are computed as follows:

$$\text{RandStatistic} = \frac{f_{00} + f_{11}}{f_{00} + f_{01} + f_{10} + f_{11}} \tag{4.24}$$

$$\text{Jaccard coeffecient} = \frac{f_{11}}{f_{01} + f_{10} + f_{11}} \tag{4.25}$$

In addition to comparing the algorithms, this technique can also be used to determine the clustering tendency of a set of data, determine the correct number of clusters, evaluate how well the results of cluster analysis fit the data without reference to external information, and compare the results of a cluster analysis to externally known results. The following section gives the pseudocode for K-means clustering and the evaluation of the results.

Ex 2: Pseudocode for K-Means Clustering and Evaluation

```
function kmeansClusterEval(dataset, labels)
%dataset - computed features for all the frames in all the audio files either
MFCC, LPC or ZCR
%labels – class labels for each observation
%size of dataset, N –no. of observations, p-dimension of data, last column
contains the class labels

[N p]=size(dataset)

x=dataset (:,[1:p-1]);
y=dataset(:, p); % class labels

%built-in function for reducing the dimensionality of data.
x = sammon(x,3);

k=unique(y) %number of clusters same as unique number of classes

%apply k-means clustering
[IDX, km_center, SUM, D]=kmeans(x, k)

%Compute the adjusted rand index using built-in function
adj_rand_index=adjrand(y,IDX);

%the function computes the purity as per the formula described in section 4,
purity is an array with purity value for all the clusters
purity=comp_purity(y, IDX);

%the function computes the purity as per the formula described in section 4
jacCoeff=comp_Jaccard(y, IDX);
```

4.4.1 *Clustering Audio Data of Small Size*

We considered a small dataset by randomly selecting the *50* observations from the dataset constructed with MFCC feature vectors of *3* types of animal voices, viz., *alligator*, *cat*, and *bat*. We applied all the clustering techniques, and for each of the clustering algorithm, the number of classes ($k = 3$) is given as input. Table 4.1 shows the purity (for each of the cluster), rand statistic, and Jaccard coefficient for all the four algorithms. The higher values for these measures indicate that clustering performance is better for the considered dataset.

The purity of all these clusters is above 85% given in Table 4.1 (first row) for MFCC. It can be observed that the evaluation measures for K-means clustering with LPC features (second row) are not as good as MFCC. The purity, rand statistic, and

Table 4.1 Purity (three clusters), Rand statistic, and Jaccard coefficient measures for clustering of animal voice data

Clustering algorithm	Feature	Purity			Rand statistic	Jaccard coefficient
		Cluster 1	Cluster 2	Cluster 3		
K-means algorithm	MFCC	0.9375	0.9375	0.8704	0.8921	0.7185
	LPC	0.5167	0.8448	0.5313	0.7289	0.4296
Fuzzy C-means algorithm	MFCC	0.9375	0.9375	0.8704	0.8921	0.7185
	LPC	0.5313	0.5167	0.8448	0.7289	0.4296
Hierarchical algorithm	MFCC	1.0000	1.0000	0.8197	0.9043	0.7488
	LPC	0.6970	0.5714	0.9259	0.7703	0.4944
GMM algorithm	MFCC	0.9574	0.8772	1.0000	0.9228	0.7908
	LPC	0.6875	0.5357	1.0000	0.7795	0.5508

Jaccard coefficient values are higher for hierarchical and GMM algorithms when compared with the K-means and fuzzy C-means algorithms. Similar experiments are done for the observations of instrument dataset by selecting *50* observations randomly from each type of *3* instruments: *flute*, *harmonium*, and *santoor* with MFCC and LPC features. The clustering techniques are applied with number of clusters, $k = 3$, as input. Table 5.2 shows the evaluation measures for the four clustering techniques with MFCC/LPC features.

It can be observed from the table that the performance of hierarchical and GMM clustering algorithms are better compared to K-means and fuzzy C-means algorithms. Since the hierarchical and GMM clustering algorithms with MFCC feature perform better, we are showing the graphical views of these results in Figs. 4.14 and 4.15. In Fig. 4.14, the class *flute* (plus symbol) is clustered into Cluster 1 (plus symbol), class *harmonium* (circle symbol) is mapped onto Cluster 3 (star symbol), and class *santoor* (star symbol) is mapped onto Cluster 2 (circle symbol). In both figures, it is observed that most of the observations in same class are also found to be in the same cluster. This point is also supported by the evaluation measure values, with purity value more than 90% for each individual cluster (Table 4.2, 5th and 7th row).

4.4.2 Clustering Large-Size Audio Data

The experiments have shown that clustering of small-size dataset is very clear with high purity, Rand statistic, and Jaccard coefficient. However, as we increase the

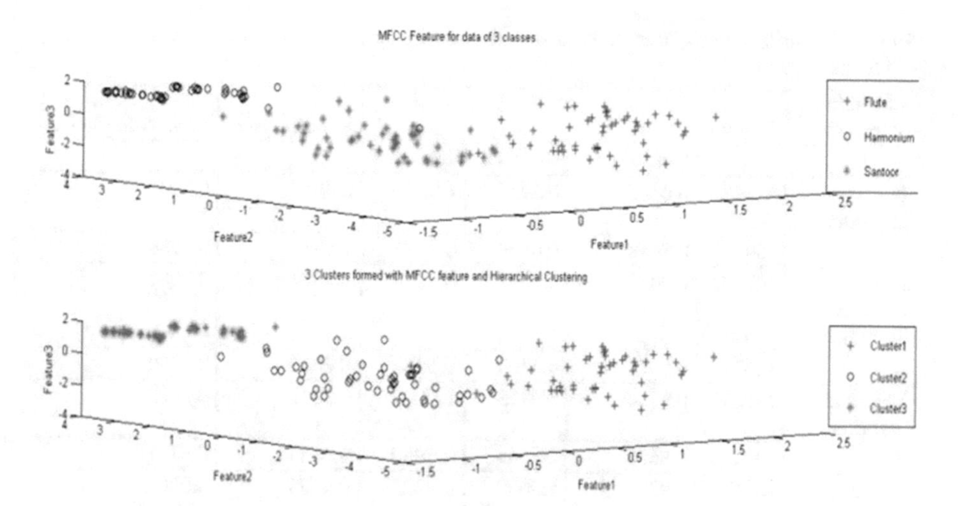

Fig. 4.14 Instrument data with MFCC feature and results of hierarchical clustering

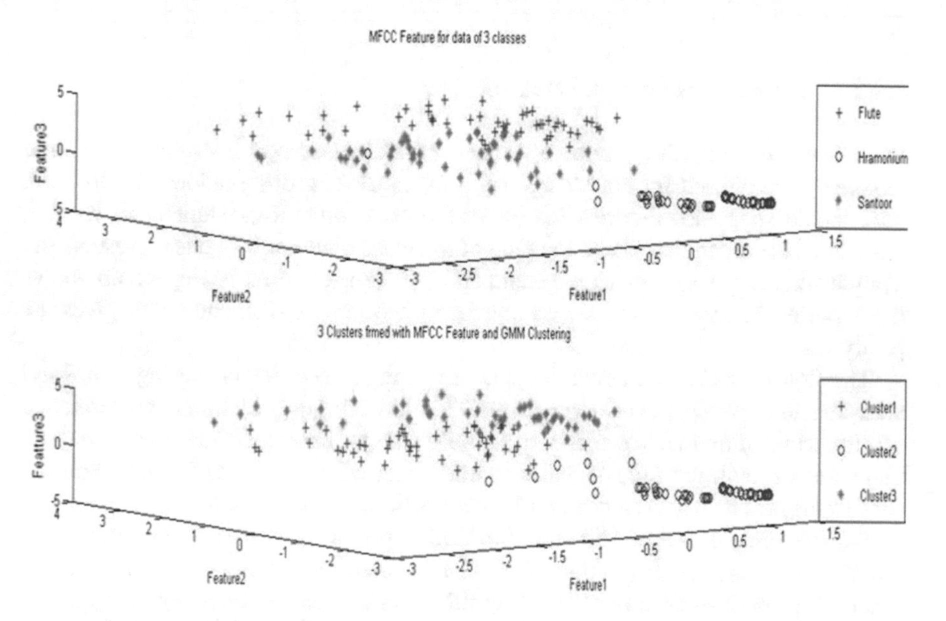

Fig. 4.15 Instrument dataset with MFCC features and results of GMM clustering

number of observations in each class, the results may vary. For analyzing this, we consider the large dataset and apply clustering techniques.

Table 4.2 Purity (three clusters), Rand statistic, and Jaccard coefficient measures for clustering of instrument data

Clustering algorithm	Feature	Purity			Rand statistic	Jaccard coefficient
		Cluster 1	Cluster 2	Cluster 3		
K-means algorithm	MFCC	0.6486	0.5938	1.0000	0.7816	0.5105
	LPC	0.5488	0.6809	0.8571	0.6813	0.3989
Fuzzy C-means algorithm	MFCC	0.9038	0.9091	0.9074	0.8874	0.7087
	LPC	0.8571	0.6809	0.5488	0.6813	0.3989
Hierarchical algorithm	MFCC	0.9800	0.9423	1.0000	0.9649	0.8988
	LPC	0.8636	0.7209	0.5059	0.6864	0.4101
GMM algorithm	MFCC	0.9792	1.0000	0.9423	0.9656	0.9007
	LPC	0.8235	0.6481	0.8222	0.7461	0.4442

4.4.2.1 Clustering of Animal Dataset

For animal dataset, *five* classes – voices of *alligator*, *cat*, *bear*, *elephant*, and *monkey* – are chosen for clustering. The ranges of total observations are *50*, *100*, *150*, and *200* per class. Figure 4.16a shows the variation of Rand statistic for varying number of observations chosen for each of the class randomly. *X*-axis represents the sample size, and *y*-axis gives the Rand statistic value. Similarly Fig. 4.16b shows the variation in purity value. *X*-axis represents the sample size, and *y*-axis gives the purity value.

The figures show that as we increase the sample size for clustering, the Rand statistic and purity values are decreasing. This indicates degradation in performance of clustering of animal voice data as sample size is increased. The reason may be that some of the frames in a signal may not represent the accurate features required for that class. The performance of all the algorithms is almost same.

For the sample size of 150 and 200, GMM performance is better compared to others. However, for the sample size of 150, hierarchical is better compared to others. The performance of all the algorithms is almost the same for sample size of 50. But, K-means and fuzzy C-means clustering are low compared to the other two algorithms. For the largest sample size, highest purity achieved is up to 50%, and the highest Rand statistic achieved is 0.2245, for five classes of animal voice data with MFCC features. Similarly, clustering algorithms are applied for the dataset with LPC features with varying sample sizes, for *five* classes of animal voices, viz., *alligator, cat, bear, elephant, and monkey.* We observe the similar variations, in the performance of the results which is shown in Fig. 4.17a, b.

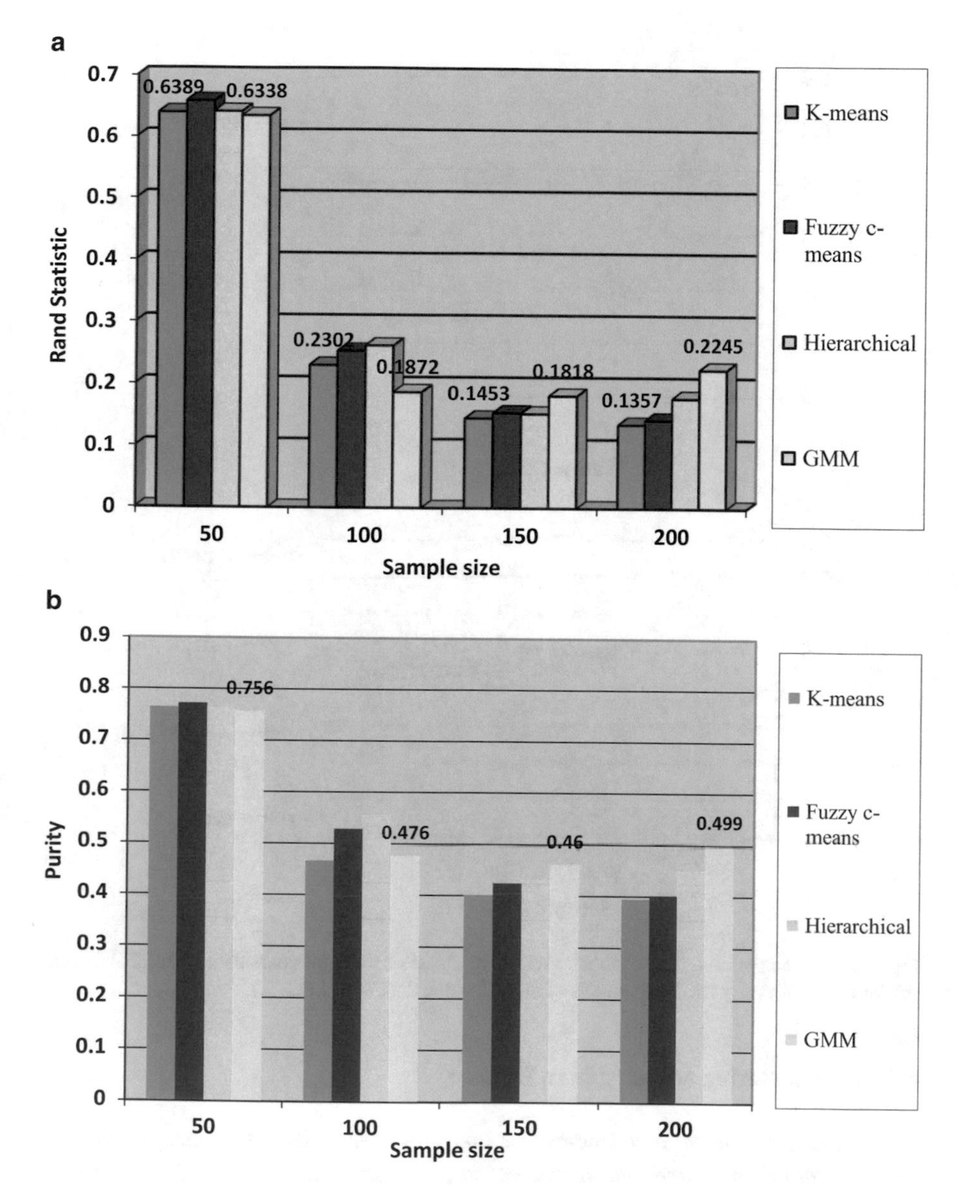

Fig. 4.16 (**a**) Variation of Rand statistic for five classes of animal voice data with MFCC. (**b**) Variation of purity for five classes of animal voice with MFCC

Even in this case, as sample size is increased, Rand statistic and purity have decreased as seen in Fig. 4.17a, b. GMM performs slightly better compared to others for the sample size of 200. For the largest sample size of 200, highest Rand statistic is 0.1612, and the highest purity is 0.47 that is low compared to that of MFCC features.

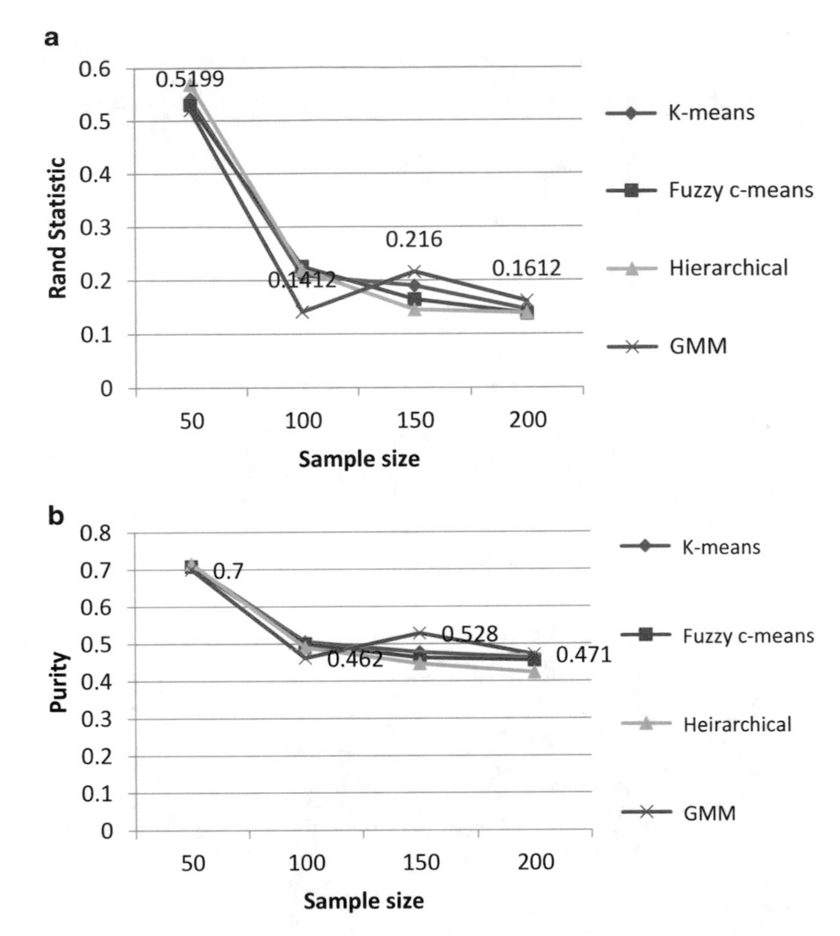

Fig. 4.17 (**a**) Variation of Rand statistic with five classes of animal voice data with LPC feature. (**b**) Variation of purity with five classes of animal voice data with LPC

4.4.2.2 Clustering of Instrument Dataset

We conducted similar experiments for instrument data with *ten* classes, namely, *flute*, *harmonium*, *mandolin*, *nagaswaram*, *santoor*, *saxophone*, *sitar*, *shehnai*, *veena*, and *violin*. The different sample sizes, i.e., 50, 100, 150, and 200, are considered to see the variations in the Rand statistic and purity value across different algorithms and features. The results of the variations in Rand statistic and the purity value are shown in Fig. 4.18a, b, respectively.

As we can observe from both figures, clustering algorithm performance is almost similar to that of the animal voice data. Even in this case, the GMM algorithm performance is slightly better compared to others. The GMM clustering result is highest for sample size of 50, 150, and 200. For the sample size of 100, fuzzy

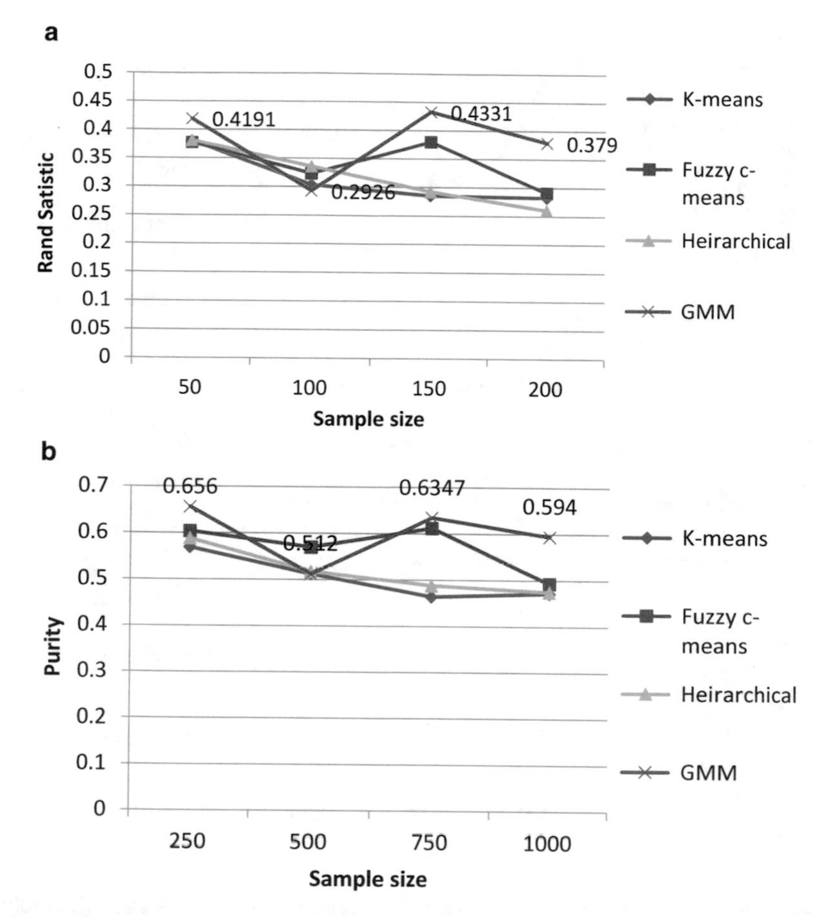

Fig. 4.18 (**a**) Variation of Rand statistic for instrument data with MFCC. (**b**) Variation of purity for instrument data with MFCC

C-means is better compared to others. Here, hierarchical and K-means algorithms perform poorly compared to GMM and fuzzy C-means algorithms. The highest Rand statistic value for the 200 observations is 0.379, and the highest purity achieved for the clustering of instrument data with MFCC feature is 59% slightly better compared to that of animal voice data.

The Rand statistic and the purity value for varying number of observations with LPC features are shown in the Fig. 4.19a, b. The highest purity achieved is 44% which is slightly less compared to the clustering with MFCC feature.

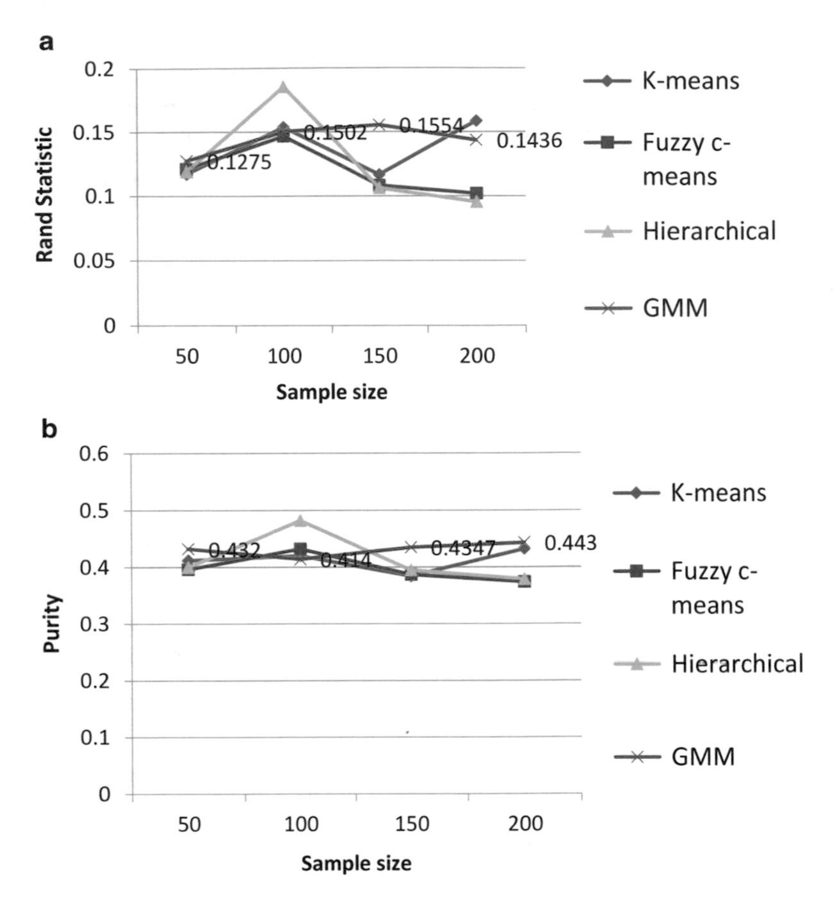

Fig. 4.19 (**a**) Variation of Rand statistic for instrument data with LPC. (**b**) Variation of purity for instrument data with LPC

4.4.2.3 Clustering of Singer Dataset

For the singer data, we considered *six* singers and conducted the same experiments as explained above. Fig. 4.20a, b show the results in the variation of Rand statistic and purity values for varying number of observations with MFCC feature.

Figure 4.21a, b below show the variation of Rand statistic and purity value for the singer dataset constructed with LPC features as we increase the number of observations.

Hierarchical algorithm is slightly better compared to other clustering algorithms in all the cases. For singer data, the highest Rand statistic achieved for MFCC feature is 0.2088, and the highest purity value is 52.3% for sample size of 200. In case of LPC feature, highest Rand statistic achieved is 0.1146 and purity is 40%.

We have also done experiments by increasing the number of clusters to be formed than the actual number of classes present. In these experiments, the purity and

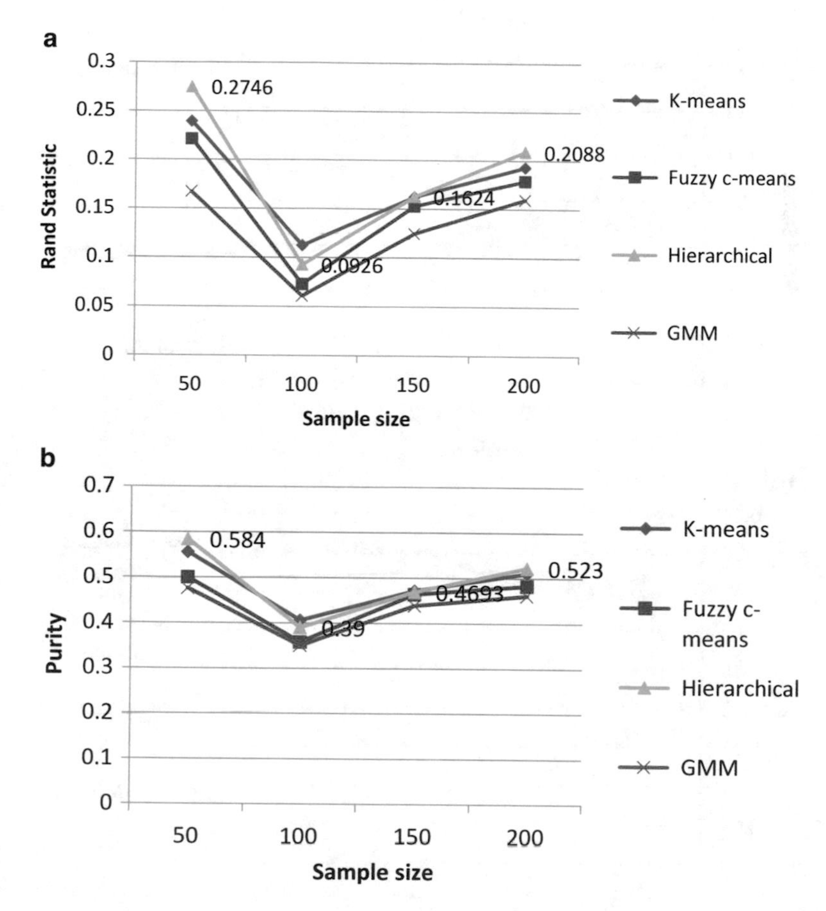

Fig. 4.20 (**a**) Variation of Rand statistic for singer data with MFCC. (**b**) Variation of purity for singer data with MFCC

entropy value increased as we increase the number of clusters to be formed. This is expected since the clusters become more homogenous.

4.5 Analysis and Conclusions

We have discussed the four types of clustering algorithms: K-means, fuzzy C-means, hierarchical, and GMM algorithms. These techniques are applied for *three* types of datasets (animal voice, instrument, and singer) across two features (MFCC and LPC). The experiments are repeated for datasets of various sample sizes taken from each class. The evaluation measures like purity, Rand statistic, and Jaccard coefficient are computed in each experiment.

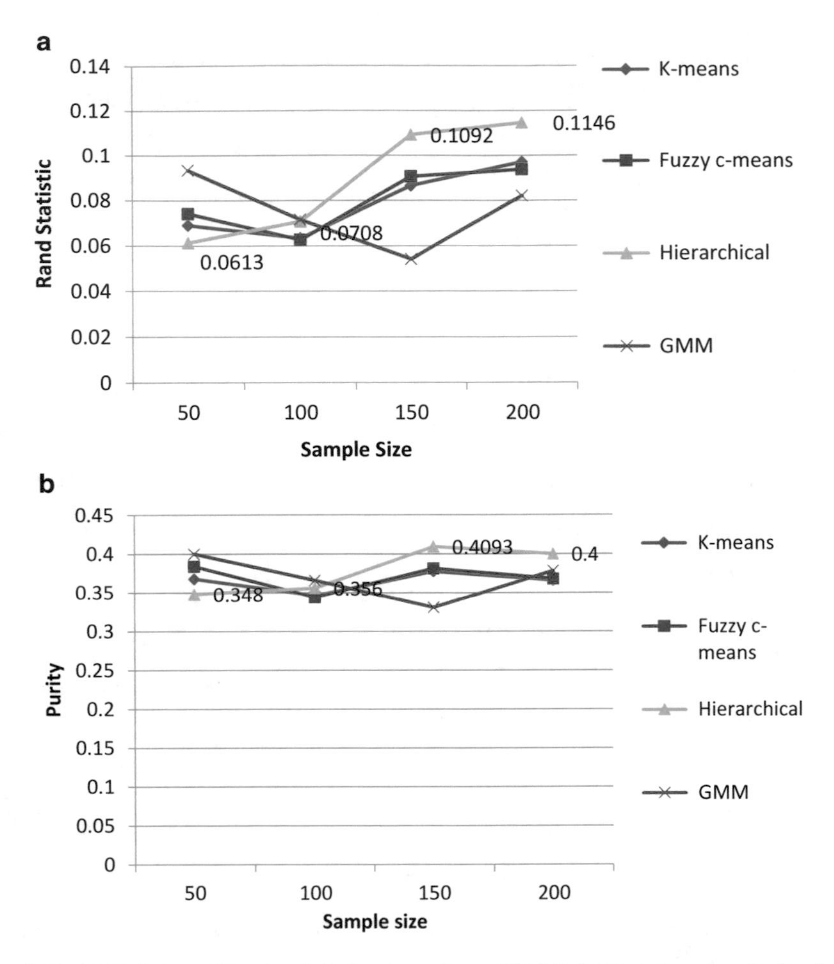

Fig. 4.21 (**a**) Variation of Rand statistic for singer data with LPC. (**b**) Variation of purity for singer data with LPC

For the small dataset of 50 observations, we found that the clusters are formed in exactly the way of classes. This is found to be true for all the dataset with all four clustering techniques. As we increased the number of observations to be clustered, the algorithms showed degradation in the performance. GMM algorithm showed better result for the sample size of 200 in case of animal voice data with both MFCC and LPC feature which is followed by hierarchical algorithm. The performance with LPC features is slightly better than MFCC for animal voice data. For instrument data, GMM algorithm clustering results are better compared to others, which is followed by fuzzy C-means algorithm. The performance of MFCC feature is slightly better compared to LPC for instrument data. The performance of all the clustering algorithms is almost same in case of singer data. Hierarchical algorithm is slightly better compared to the other algorithms. In this case also, MFCC feature is better than LPC for the sample size of 200.

The results show that there exists a natural structure within the data of same class in the given audio data. This is clear when we take small-size data and apply clustering techniques. But as we increase the size of dataset, clustering algorithm could not detect the natural structure within the data very efficiently. This type of analysis would be helpful for comparing the results of clustering algorithm for audio datasets. It also determines which kind of feature demonstrates natural structure better than others for a particular type of dataset.

References

1. Alpaydin E (2004) Introduction to machine learning. PHI Learning, Delhi ISBN-81-203-2791-8
2. Pachet F, Aucouturier JJ (2004) Improving timbre similarity: how high is the sky. J Negat Results Speech Audio Sci 1(1):1–13
3. Casey MA, Veltkamp R, Goto M, Leman M, Rhodes C, Slaney M (2008) Content-based music information retrieval: current directions and future challenges. Proc IEEE 96(4):668–696
4. Duda RO, Hart PE, Stork DG (2006) Pattern classification. Wiley-Interscience Publication, USA
5. Dunham MH (2003) Data mining introductory and advanced topics. Pearson Education Inc, India ISBN: 81-7808-996-3
6. Gold B, Morgan N (2006) Speech and audio signal processing – processing and perception of speech and music. Wiley India Pvt Ltd, India ISBN: 81-265-0822-1
7. Martinez WL, Martinez AR (2008) Computational statistics handbook with Matlab. Chapman & Hall/CRC Publications, USA ISBN: 1-58488-566-1
8. McLoughlin I (2009) Applied Speech and Audio processing with MATLAB examples. Cambridge University Press, USA
9. Peltonen V, Tuomi J, Klapuri A, Huopaniemi J, Sorsa T (2002) Computational auditory scene recognition. In: IEEE international conference on acoustics speech and signal processing, vol 2
10. Rabiner L, Jaung B (1993) Fundamentals of speech recognition. Prentice Hall, USA ISBN,-10:013051572
11. Shin K, Hammond JK (2008) Fundamental of signal processing for sound and vibration engineers. John Wiley & Sons Ltd, England
12. Slaney M (1998) Auditory toolbox: a MATLAB toolbox for auditory modeling work, Tech. Rep. 1998–010. Interval Research Corporation, Palo Alto Version 2
13. Smith SW (1997) The scientist and engineer's guide to digital signal processing. California Technical Publishing, San Diego, California www.DSPguide.com, ISBN 0-9660176-6-8
14. Tan P, Steinbach M, Kumar V (2006) Introduction to data mining. Pearson Addison Wesley, USA ISBN: 978-81-317-1472-0
15. Theodoridis S, Pikrakis A, Koutroumbas K, Cavouras D (2010) Introduction to pattern recognition: a matlab approach. Academic Press, USA
16. Michael W (2007) Digital signal processing using MATLAB and wavelets. Infinity Science Press LLC, Hingham
17. Xu C, Maddage NC, Shao X, Cao F, Tian Q (2003) Musical genre classification using support vector machines. In: Proceedings of IEEE international conference on acoustics, speech and signal processing (ICASSP'03), vol 5, Hong Kong, pp 429–432

Chapter 5
Feature Extraction and Classification in Brain-Computer Interfacing: Future Research Issues and Challenges

Debashis Das Chakladar and Sanjay Chakraborty

5.1 Introduction

Brain-computer interface (BCI) is a useful communication system for paralyzed people with the neuromuscular disorder that does not require any peripheral muscular activity. Brain signal can be captured by several electrodes, and scalp voltage is measured in a timely manner. Different scalp voltage level of brain signal is transferred into cursor position. An efficient way of cursor movement to the target is described in the paper [25]. In paper [68], the author has described a powerful classifier that combines all the user response and outperforms conventional motor imagery classifier. In paper [69], the author has illustrated the interaction between affective phenomena of human behavior like moods and emotions in affective BCI (aBCI). He also described some applications and challenges of aBCI. However, in BCI various brain activity patterns of disabled people are identified and translated into necessary commands with the help of various classification and signal processing algorithms. BCI has several application areas like wheelchair control, spelling device, robot control, etc. There are three approaches (invasive, partially invasive, and noninvasive) which exist in BCI based on electrode position in the brain. In this survey, we are focusing on EEG signal which belongs to the noninvasive unit. Electric brain signals are captured by different electrodes, and that signal is classified by EEG [80]. Then the useful features are extracted from that signal, and those features are classified accordingly. Finally, the signal is transferred to application interface where that signal is converted into proper action for disabled

D. D. Chakladar (✉)
Computer Science and Engineering Department, Institute of Engineering and Management, Kolkata, India

S. Chakraborty
Department of Information Technology, Techno India, Kolkata, India

X. Li, K.-C. Wong (eds.), *Natural Computing for Unsupervised Learning*, Unsupervised and Semi-Supervised Learning,
https://doi.org/10.1007/978-3-319-98566-4_5

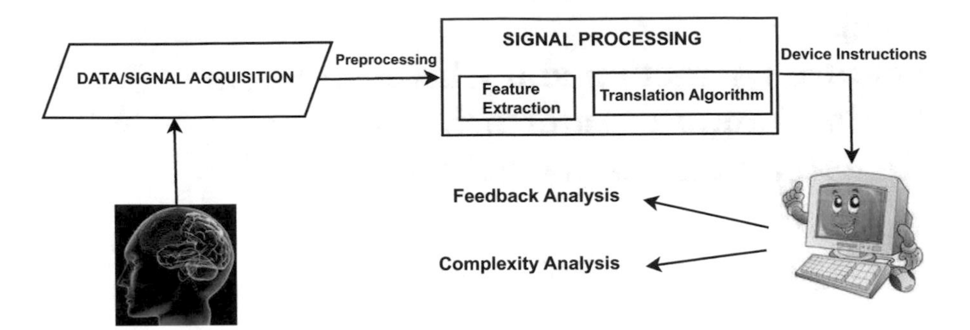

Fig. 5.1 Basic architecture of a BCI system [25]

people. Several works have been done on feature classification process in BCI due to the rapidly growing interest for EEG-based BCI. Very few review works [66, 101] on classification algorithms for BCI have been published till date. In this survey, we have discussed all the important classifiers (linear, nonlinear) used in EEG-based BCI systems and perform a detailed comparative analysis after identifying their critical properties. Based on the comparison, we can guide the researchers about the best classifier in terms of performance for a given BCI experiment. The basic architecture of a BCI system is shown in Fig. 5.1.

5.2 Outline of the Survey

In this paper, we describe basic introductory concepts of a BCI system in Sect. 5.1. We focused on different segments of BCI technology (especially feature extraction and selection) in Sect. 5.3 and also explained each of the feature extraction techniques used in BCI in details with their applications. We analyze each of the classifier performance for different BCI tasks in Sect. 5.4. In Sect. 5.5, we discuss some open problems related to EEG-based BCI classification techniques which may guide the interested researchers in their future work. Finally, we have concluded and assess their usability in Sect. 5.6.

5.3 EEG-Based Brain-Computer Interfacing Through Feature Classification

5.3.1 Signal Acquisition

EEG signal is a measurement of current that flows during synaptic excitation of the dendrites. When the brain cells (neurons) are excited (ON state), then synaptic current is produced within the dendrites of those cells. The current produces electric field over the scalp, and then it transfers throughout the body [102]. Either internal noise (produced within the brain) or external noise (produced over the scalp) is generated during signal acquisition. The noisy signal may reduce the signal's strength so that reduced signal is captured by the electrodes [86] which are used in later stages. Here we have described the classification of EEG signal and different types of brain signal used in BCI, respectively.

5.3.1.1 Classification of EEG Signal

1. *Alpha rhythm:* It describes the physical and mental rest. It has a high frequency (8–13 Hz) and low voltage [84].
2. *Beta rhythm:* It describes the state of consciousness with high frequency (13–20 Hz) [84].
3. *Theta rhythm:* It describes the deep and normal sleep with low frequency (4–8 Hz) [84].
4. *Delta rhythm:* It describes the pathological states of neuronal difficulty(comma) with low frequency (0–4 Hz) [84].
5. *Gamma rhythm:* It is the brain wave usually associated with activity excitement with very high frequency (21–30 Hz) [84].

5.3.1.2 Types of Signal

1. *P300:* This signal is the component of ERP (event-related potential). ERP response is generated after a particular external or internal event occurs.
2. *SSVEP:* This signal has been generated in response to a visual stimulus. Nowadays, many robotic control visualizations have implemented using this signal [82].The drawback of this technique is that often paralyzed people are unable to move their eyes using this signal.
3. *Sensorimotor rhythm(SMR):* This rhythm generates at frequency level 8–12 Hz [27]. SMR can detect the current and future prospects of BCI in clinical applications [105].
4. *Slow cortical potential (SCP):* It is the slow cortical potential caused by the shift in depolarization stage of dendrites. Negative SCP means the sum of synchronized potential caused in a certain time range [82].

5.3.2 Signal Processing

Brain signal acquired from electrodes may contain noise, so the noise needs to be removed for feature recognition and classification. The signal-to-noise ratio (SNR) is a measurement of signal and noise power. A high SNR means minimal corruption of signal due to noise. Noise is generated due to background neurological activity and artifacts which are produced from biological and external sources [101]. Signal processing has several sections by which the actual digital signal is passed to the translation algorithm after removing noise and artifacts.

5.3.3 Feature Extraction

5.3.3.1 Properties of Feature

There are many features that are related to the BCI design such as amplitude value of EEG signals, band powers (BP), and autoregressive(AR) and adaptive autoregressive (AAR) parameters [64, 87]. Concerning the design of the BCI system, some important properties of these features are described below.

1. *Noise and outliers:* BCI signals are noisy in nature as EEG signal has low signal-to-noise ratio.
2. *High dimensionality:* In BCI, feature vectors are high dimensional in nature, so several features are extracted from the brain in different time range and merged into a single feature vector.
3. *Time information:* BCI features must contain the time information, as the signal pattern of the human brain has changed time to time.
4. *Small training sets:* A training set of BCI is relatively small since the training process is time-consuming, so taking small training data takes minimum computational time.

5.3.3.2 Feature Extraction Using Fourier Transform

EEG signal is classified with respect to time and frequency levels. But the signal features are exploited by unwanted noise. So in order to extract useful features from the EEG signal, fast Fourier transform (FFT) is applied over EEG signal to get the corresponding useful spectrum [31]. For nonstationary nature of the EEG signal, FFT cannot be used directly, so short-time Fourier transform (SSFT) is used to extract useful features from the EEG signal. In paper [6], the author has described the time-frequency representations (TFRs) of FFT. In paper [97], some other types of feature generation methods from the nonlinear system have been discussed. Features are created based on some proposed mechanism, and it is discussed in the papers [33, 34, 40]. The drawbacks of this method have been described in the paper [96].

5.3.3.3 Feature Extraction Model

Different thinking activities produce different patterns in the human brain. BCI is treated as a pattern recognition system by classifying each pattern into appropriate class based on the useful features. Brain signals are generally nonstationary in nature. A good technique called stationary subspace analysis (SSA) is implemented to deal with nonstationary brain signals [99]. The following table describes different feature extraction categories along with their implementation details (Table 5.1).

Table 5.1 Summary of feature extraction techniques

Category	Method	Properties	Applications
Dimension reduction	PCA	Linear transformation	[17, 62, 66]
		Set of correlated observations are transformed into set of uncorrelated variables	
	ICA	Valuable noise and dimension reduction method	
		Remove signal processing time and increase data communication rate	[37]
		Decompose EEG signal into different independent components useful for artifact removal	[21, 60]
Space	CSP	Maximizes the difference between the variance of data density distribution of two classes	[67]
		Increases the performance of synchronous BCIs but returns poor results in asynchronous BCIs	[20]
		Improved version of CSP – wavelet common spatial pattern (WCSP)	[20]
Time-frequency	AR	Useful for different applications like data compression and classification	[58]
		Provides compact and efficient representation of EEG signal	[48]
		Performs more consistently for non-segmented data	
		Improved version of AR: multivariate adaptive AR (MVAAR)	[16]
	CWT	Useful for event-related brain potential signal	[24]
	DWT	Provides both frequency and temporal information of the signal	[47]
		Reduces the redundancy and complexity of CWT	[54]
		Divides the EEG signal into overlapping windows and performs classification	[46]
		DWT and modified fractal dimension method extract features from multiscale data	[57]

1. *Principal component analysis (PCA)*
   PCA is a method for extracting features after reducing the dimension of data. PCA projects the input data on a k-dimension eigenspace of k eigenvectors. All the data points (X) are projected in the direction of the first eigenvector (v) so that resultant variance (Z) will be at maximum. Significant eigenvalues are selected using ANOVA test, which is then trained and tested using several supervised classifiers [57]. PCA also removes artifacts for reducing the dimension of the signal [2].
2. *Independent component analysis (ICA)*
   ICA is another popular method of extracting features. ICA assumes that the observed EEG signal is a mixture of several independent source signals. It expresses the resulting signal *x(t)* with respect to source signal *s(t)* in the following way.

$$x(t) = f(s(t)) + n(t) \tag{5.1}$$

   where f is mixer function, *n(t)* is noisy signal, *s(t)* is source signal, and *x(t)* is output signal. Dimension of the source signal depends on the number of sources, and dimension of the output signal depends on the number of measured data channels [74].
3. *Common spatial pattern (CSP)*
   CSP is a feature extraction method that projects multichannel EEG signals into a subspace, where the differences between classes are highlighted and the similarities are minimized. CSP can be regularized into stationary subspace (SCSP), and it increases the classification accuracy [85]. In current days, CSP has been implemented in multi-class-level classification based on some heuristic approach [42].
4. *Autoregressive component (AR)*
   AR method is useful for feature extraction with respect to the time domain. AR model uses parametric approach to get better frequency without any problem of spectral leakage; hence it performs better than nonparametric approach [7]. AR spectral estimation prefers to use Fourier transform, but it performs poorly in nonstationary signal [56]. To overcome this problem, multivariate adaptive AR (MVAAR) has implemented [76, 100].
5. *Continuous wavelet transform (CWT)*
   CWT is a powerful technique to extract valuable features from the signal, and it is also used as a pattern recognition tool [8]. Using some properties of wavelet template, CWT performs better than classic template-matching method. The wavelets are suitable for transient signal analysis, where spectral properties of the signal change in time [26].
6. *Discrete wavelet transform (DWT)*
   The main drawback of CWT is that it produces redundancy and complexity as it involves the analysis of a signal at a very high number of frequencies after shifting of mother wavelet [74]. DWT translates and dilates the mother wavelet in discrete values [26]. The multi-resolution of raw EEG data *x(n)* has been divided into two digital filters *g(n)* and *h(n)* and divides the sample by two [23]. The

discrete mother wavelet *g(n)* is a high-pass filter, whereas the mirror image *h(n)* is a low-pass filter [26].

5.3.4 Feature Selection

In order to select an appropriate classifier for a given BCI system, it is necessary to understand important features, their properties, and their relationship. BCI input generally deals with high volume of features, such as amplitude values of EEG signals, band powers (BP), power spectral density (PSD) values, autoregressive (AR) and adaptive autoregressive (AAR) parameters, time-frequency features, and so on. Feature selection methods help the researchers to choose and study significant features from those high volumes of features to execute the entire process with ease. Feature selection process is quite different from extraction. Feature selection process helps researchers to identify important feature subset which has been extracted through various methods of feature extraction. A good feature selection and extraction process also help to select an appropriate classifier. In most of the cases, feature selection plays a vital role in EEG-based signal classification. The selection of a subset of features that are most useful to the classification problem often increases classification accuracy on new data [39]. In the paper [39], genetic algorithm (GA) is used to search the space for feature subsets, and SVM classifier is used for classifying the data. GA evaluates many points in the space in parallel and provides a stochastic global search of the feature subset space. In the paper [77], a wrapper approach is used for feature subset selection to maximize the classification (SVM) accuracy in EEG-based BCI research. This accuracy speed is necessary because we will train and test the classifiers for every feature subset we evaluate.

5.3.5 Feature Classification

After the required features have been extracted from the EEG signal, those useful features need to be classified appropriately using a suitable classifier. Feature classification method has two parts, (1) regression and (2) classification, but classification approach is the most suitable one used in recent days [65, 87]. Regression algorithms are useful for quantitative inputs while classification algorithms mainly handled categorical inputs. Regression algorithms are mostly used when we try to predict the user's intention based on the extracted features from EEG signal, whereas classification algorithms are used to classify the independent features to the appropriate class. Classification algorithms are mostly used for two class problems as it is sometimes called "binary classification." Therefore, classification approach might be more useful for two-target applications. In paper [64], the author has described that regression algorithms are better for multiple targets. Here we have discussed several feature classification categories including linear, nonlinear, and graphical. We also briefly discuss important properties and application of several approaches/models which belong to that category (Table 5.2).

Table 5.2 Summary of different feature classification methods

Taxonomy	Approach	Properties	Applications
Generative model	Bayesian analysis	Uses prior statistical knowledge of the feature and parameter estimation technique (maximum likelihood) for finding the posterior output	[78]
		Not very popular in the BCI systems	
Linear	LDA	Draws a hyperplane to separate features into the different classes based on discriminant function	[30]
		Fails in the presence of strong noise and high dimensional feature	[39]
		Classification of event-related potential (ERP) signals of the brain	[15]
		Improved versions of LDA – BLDA and FLDA	[9, 98]
	SVM	Maximizes the distance between the nearest training data points and the separating hyperplanes	
		Selecting optimal number of channels varies with some constant	[81]
		Useful for linear and nonlinear approach	[70]
		Gives poor results in the presence of outliers, so regularization required	[38]
		Speedy classifier	[95]
Nonlinear	KNN	Uses metric distances between features of the testing data and their neighbors	[49]
		Gives good result if number of feature vectors are small and locally distributed	[52]
		Gives high performance in pattern recognition problems	[10]
	ANN	Simple approach, returns high level of accuracy in testing	[49]
		Gives more accurate classification with PCA	[55]
		Not suitable for unbalanced class distribution	[104]
		Advanced model of ANN – PNN	[43]
	QDA	Uses different covariance matrix for each class	
		Useful for left and right limb movements in BCI	[13]
Graphical methods	HMM	Useful for online classification	[75]
		Applicable in pointing movements and mental task classification	[51, 72]

5.3.6 Classification Methods

5.3.6.1 Bayesian Statistical Classifier

This classifier is mainly used in wheelchair steering application in BCI. A P300-based visual stimulus has controlled the direction of selected target for steering the wheelchair. The classification uses the prior probability of target and nontarget components and computes the posterior probability of targets to move the wheelchair [78]. For posterior probability, it is assumed that distribution of input features in each class is Gaussian, but sometimes the distribution is a weighted mixture of different Gaussian components [103]. This classifier also used an adaptive variation Bayesian classifier when the EEG segments translate into probabilities of cognitive states to control different mental states [93].

5.3.6.2 Linear Discriminant Analysis (LDA)

LDA gives good accuracy in different BCI systems like P300 speller [16], multi-class BCI [79], etc. Improved versions of LDA such as Fisher LDA (FLDA) and Bayesian LDA (BLDA) have been implemented where the first one is useful in projecting the data at lower dimensional space and the second one is useful when the number of features is very large [9, 98]. FLDA and BLDA give high classification accuracy for disabled and able-bodied subjects in BCI. FLDA- or BLDA-based system is used for the paralyzed people [9], as BCI provides them better communication and control to restore their lost abilities [106].

5.3.6.3 Support Vector Machine (SVM)

SVM uses optimal hyperplane to separate feature vectors between two classes. Different type of kernel functions is available for feature classification. The Gaussian kernel function in SVM is applied in BCIs to classify P300-evoked potentials [106]. SVM is implemented in multi-class channel analysis of EEG signal and evaluated using Cohen's kappa coefficient [88]. In paper [61], the author has described that semi-supervised SVM has significantly reduced the training effort of P300-based BCI speller.

5.3.6.4 K-Nearest Neighbor Classifier (KNN Classifier)

In KNN, the predicted output of input feature vector is the average output value of all the "K" neighbors within the feature space [35]. KNN is used to classify the spectrogram image using brain wave balancing application in BCI [71]. It is also used to classify normal brain activity through EEG signal. This algorithm was used to classify chronic mental stress and performs satisfactory accuracy, sensitivity analysis [53]. KNN algorithm is also used to classify the features of left- and right-hand motor imagery from the EEG signals [10]. Sentiment analysis to predict future trends through feature selection using naive Bayes theorem, SVM, and KNN classifiers is well discussed in the papers [19, 28].

5.3.6.5 Artificial Neural Network (ANN)

This algorithm follows human brain processing methodology. This model is heavily used in pattern recognition as this approach learns from training data. ANN gives high classification accuracy for different mental and motor tasks [92]. Advanced approaches of ANN (APNN and PNN) have been used to classify time-varying EEG signal in BCI [43, 63].

5.3.6.6 Quadratic Discriminant Analysis (QDA)

This method is useful in left and right limb movement classification of EEG signal [13]. In paper [13], the author has implemented that QDA gives more accuracy to find the wavelet coefficients than the other classifiers (LDA and KNN).

5.3.6.7 Hidden Markov Model (HMM)

It is simply a dynamic Bayesian network vastly used in speech recognition [59]. HMM is useful for online classification of EEG data during imagination of a left- or right-hand movement [75]. HMM is also used to classify ERD and ERS features of EEG signal for finger movements [94].

5.3.7 Combination of Classifiers: Ensemble Methods

The ensemble classifier gives better accuracy in BCI P300 speller and online BCI applications [36]. This method also reduces the misclassification error and increases the accuracy of EEG signal [4]. Ensemble methods consist of three techniques – bagging, boosting, and random forest.

5.3.7.1 Bagging

In paper [90], the author has described that the enhanced version of bagging (random under-sampling bagging) with SVM can improve the performance of P300-based BCI by solving the data imbalance problem in EEG dataset. This technique significantly reduces the variance of training samples and increases the prediction accuracy of the output data [50].

5.3.7.2 Boosting

The gradient-boosting algorithm is used to detect P300-evoked potential from the EEG signal and builds several BCI applications. The classification accuracy of the gradient boosting is very high, and its model complexity is much less than other ensemble classifiers [45]. It performs poorly in the presence of noise in the feature vectors but performs well in less noise [89].

5.3.7.3 Random Forest

In this method, random selection of feature subspace is used to construct individual classifiers and build separate trees. The output from different classifiers is combined together by using the majority vote to give the final prediction [29]. In paper [5], the author has described an efficient word-typing P300 system using modified T9 interface and random forest classifier, which improves the classification accuracy by combining multiple decision trees.

5.3.7.4 Other Feature Selection Methods

1. *Evolutionary algorithm*
   The evolutionary algorithm has greater accuracy than other filtering methods [3]. Different approaches have been used for feature selection using evolutionary algorithm in BCI, like hill climber, genetic algorithms, iterated local search, and memetic algorithms [3]. But GA performs vastly in BCI applications than other evolutionary methods.

5.3.7.5 Genetic Algorithm (GA)

In paper [22], the author has described several feature extraction methods (CWT, DWT, AR) and then applied the genetic algorithm to select the subset of features that discriminate two classes of motor imagery for the EEG training data. The author also suggests that GA improves the classification accuracy based on kappa coefficient. Hybrid GA algorithm (GA followed by local search) improves the performance of LF-ASD-based BCI. This algorithm customizes the energy normalization of asynchronous BCI and gives better classification accuracy than normal GA algorithm [32].

2. *Interval type-2 fuzzy logic systems (IT2FLS)*
   For noisy and nonlinear characteristics of the EEG signal, Thanh Nguyen (author of IT2FLS [73]) has implemented the fuzzy logic-based classifier to classify the signal properly. Motor imagery signal has been classified using IT2FLS with

the help of wavelet coefficients. This classification method is better than SVM, ANN, KNN, and boosting algorithms [73]. IT2FLS is also designed to classify left- and right-hand movement based on time-frequency information extracted from the EEG signal. IT2FLS handles uncertain EEG data efficiently, and it is more robust to noise [44].

5.4 Experimental Analysis of Relevant Literature

5.4.1 Comparative Analysis Between Basic Classifiers

5.4.1.1 Comparative Analysis Between FLDA, BLDA, and NN

In paper [9], the author has described the classification accuracy between neural networks (NN), FLDA, and BLDA in P300-based BCI system using a different number of electrodes. The experiment is done on some disabled and able-bodied persons. The P300-evoked potential of the brain has been captured in the electrodes and then extracted useful features from the signal. Then those useful features have been classified using different classifiers (FLDA, BLDA, NN). Fig. 5.2 illustrates the classification accuracy between FLDA, BLDA, and NN classifiers for the different number of electrodes in P300-based BCI system [9].

From Fig. 5.2, we can conclude that BLDA has higher accuracy than other two classifiers (FLDA, NN). FLDA has suffered from overfitting problem when the number of features (p) is larger than the number of inputs (n). Using a large number of electrodes in the experiment also increased the noise or artifacts. Neural network (NN) gives the worst result due to its inherent model complexity. Due to the removal of overfitting problem, the error rates of BLDA are less than all other classifiers (Fig. 5.3) for a large number of data points.

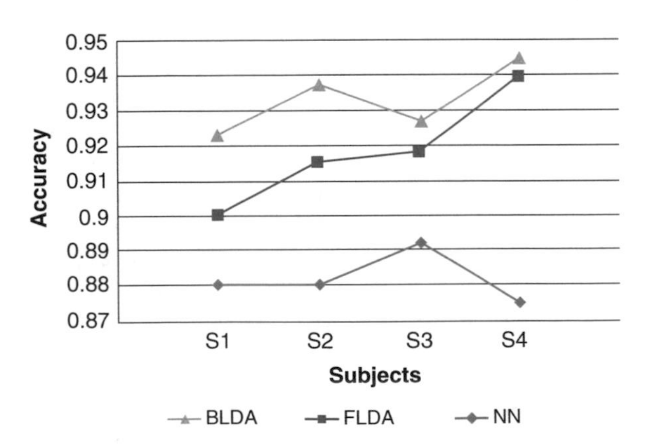

Fig. 5.2 Classification accuracy between NN, FLDA, and BLDA with 4, 8, 16, and 32 electrodes

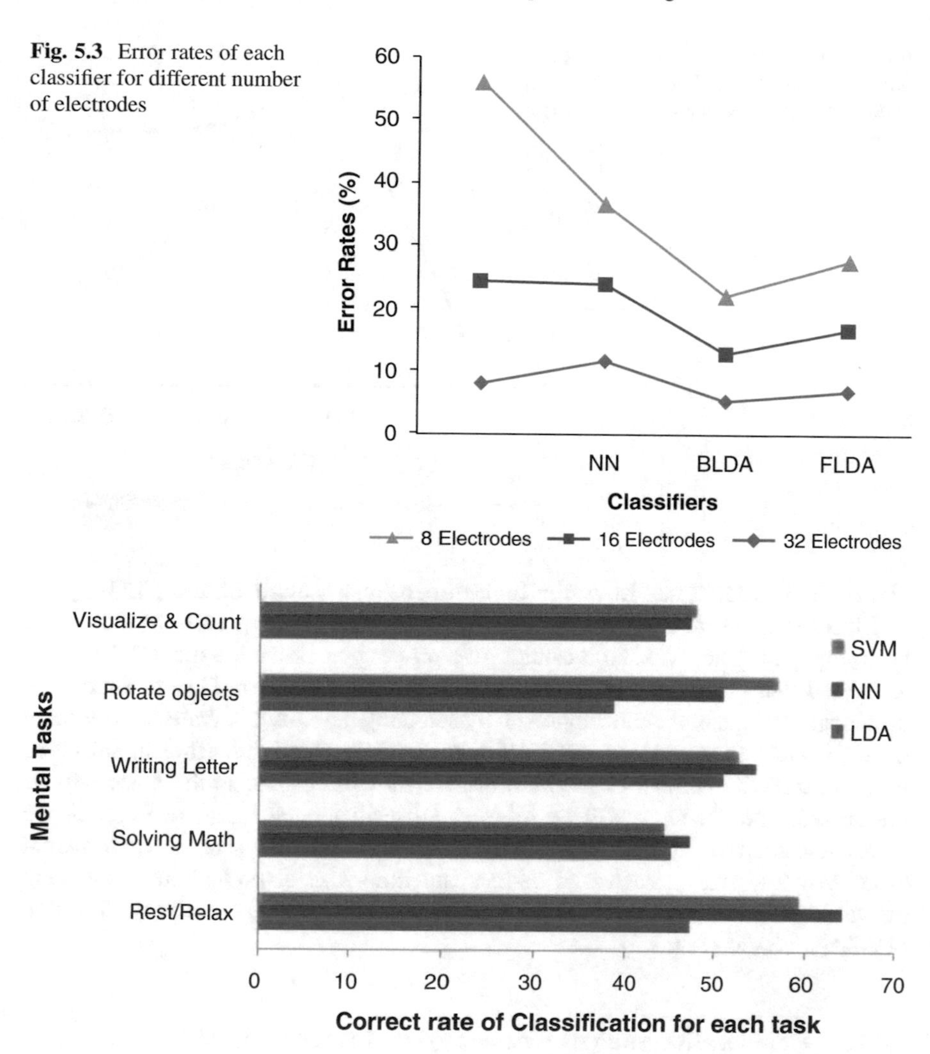

Fig. 5.3 Error rates of each classifier for different number of electrodes

Fig. 5.4 Accuracy analysis between LDA, NN, and SVM based on five different mental tasks

5.4.1.2 Comparative Analysis Between SVM, LDA, and NN

In paper [39], the author has compared linear (LDA) vs nonlinear (SVM, NN) classifiers to classify the EEG signal for five different mental tasks. Brain signal for each task (relaxing, writing a letter, solving puzzles, counting, and rotating objects) has been taken for some seconds, and each task is repeated five times. The comparison between linear (LDA) and nonlinear (NN, SVM) classifiers based on five different mental tasks is described below.

From Fig. 5.4, we can conclude that nonlinear classifier (SVM) gives better classification result than linear classifier (LDA). Nonlinear classifier specially SVM

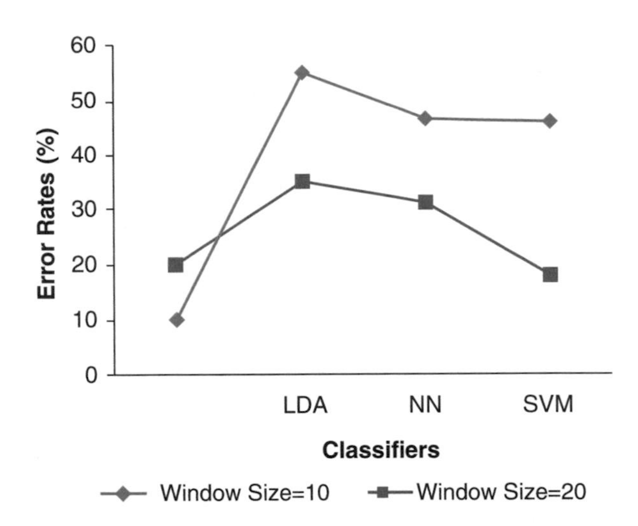

Fig. 5.5 Error rates of different classifier based on window size per trial in the experiment

gives better result than LDA for high-dimensional nature of the EEG signal. NN has the problem of choosing a proper number of hidden units in multilayer backpropagation network, so it cannot give proper generalization model. SVM has recovered that problem with the help of the genetic algorithm. Genetic algorithm provides better generalization approach by selecting the subset of features efficiently from the large feature space. As SVM can choose a small and efficient subset of features, then the number of required electrodes will be less, so the noise will be minimized, and accuracy will be improved for all mental tasks. In Fig. 5.5, we calculate error rates of three classifiers for different window sizes in each mental tasks. Window size in each trial defines the time-wise classification of different mental tasks. Error rates of SVM are less than other classifiers due to efficient use of GA with SVM (Fig. 5.5).

5.4.1.3 Comparative Analysis Between QDA, LDA, and KNN

In paper [13], the author has described the performance analysis of LDA, QDA, and KNN to classify left and right limb movement from the EEG signal. At first, the EEG signal is processed, and then useful features (wavelet coefficient, PSD, average band power, etc.) are extracted from that signal. Then these features are individually fed into different classifiers. Sometimes the features are combined into a group and passed to the classifiers. After the feature extraction phase, different classifiers (LDA, QDA, KNN) can classify the left and right limb movement from the noise-free EEG signal. The classification accuracy of each classifier based on the specific feature is described below.

From the figures (Figs. 5.6, 5.7, and 5.8), we can conclude that wavelet coefficient performs best with QDA, and in the case of PSD coefficient, both QDA and KNN give equally the highest accuracy. But KNN beats all others when average band

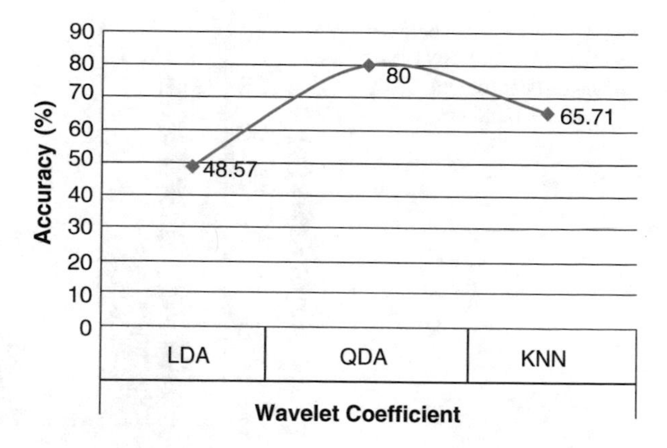

Fig. 5.6 Accuracy analysis between QDA, LDA, and KNN with wavelet coefficient

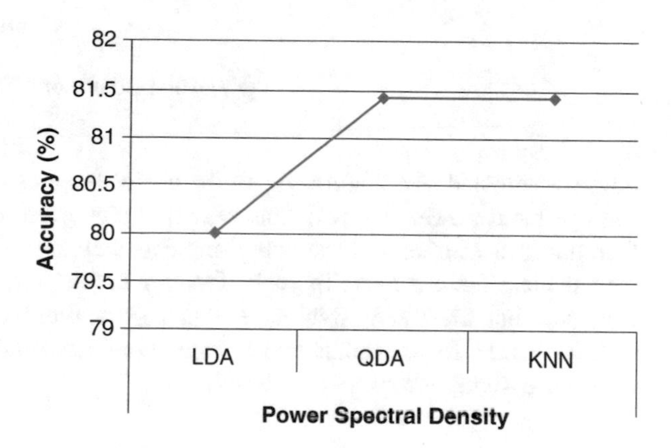

Fig. 5.7 Accuracy analysis between QDA, LDA, and KNN with PSD coefficient

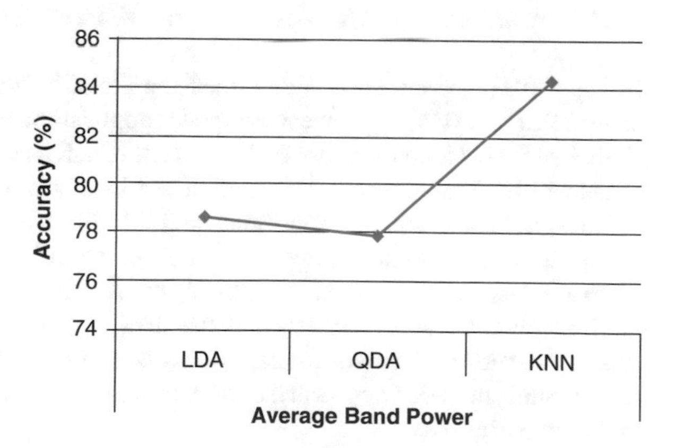

Fig. 5.8 Accuracy analysis between QDA, LDA, and KNN with average band power coefficient

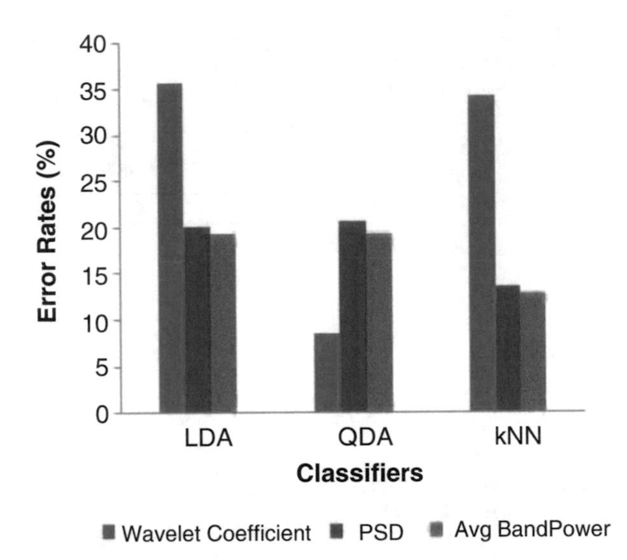

Fig. 5.9 Error rates between QDA, LDA, and KNN based on wavelet coefficient, PSD, and average band power

power is selected as a feature. As all the useful features (wavelet coefficient, PSD, average band power) support nonlinearity, QDA gives overall high performance than linear classifiers for classifying single as well as multiple features. In Fig. 5.9, we calculate the error rate of each classifier for the three mentioned features and conclude that LDA and KNN have higher error rate for wavelet coefficient than QDA, whereas for the other two features (average band power, PSD), LDA and QDA have higher error rate than KNN.

5.4.1.4 Comparative Analysis Between SVM and HMM

In paper [91], the author has described the classification of NIRS signal in BCI using SVM and HMM to drive word speller application. Near-infrared spectroscopy (NIRS) signal is a noninvasive BCI approach which uses light in the infrared range to detect blood flow, metabolic status of the localized area within the brain. Here NIRS signal is captured from the brain, and then the useful features are extracted for finger tapping and motor imagery tasks. Then classifiers (HMM, SVM) match the patterns of extracted features and classify the signal. The classifier tries to produce the best classification result so that the user can select the left- and right-hand imagery to move the cursor for selecting a box in the word speller application. We have plotted the accuracy analysis of each classifier based on finger tapping and motor imagery tasks.

From the figures (Figs. 5.10 and 5.11), we can conclude that HMM model can outperform SVM in both the tasks based on NIRS signal. The experiment is done for each pattern (left- and right-hand motor imagery and finger tapping) using HMM and SVM. As the HMM model is based on probabilistic model and it performs well

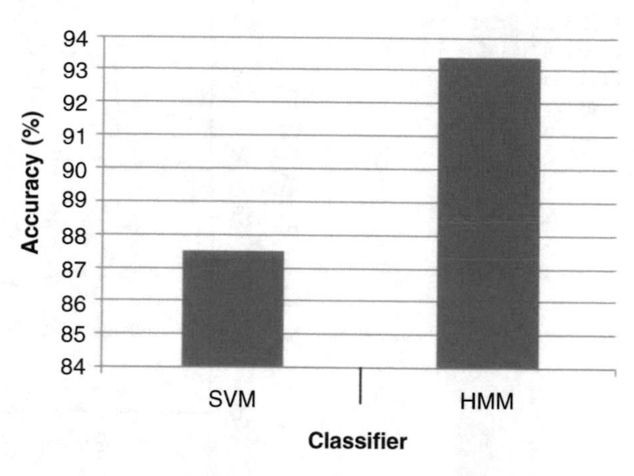

Fig. 5.10 Accuracy analysis between HMM and SVM for finger tapping tasks

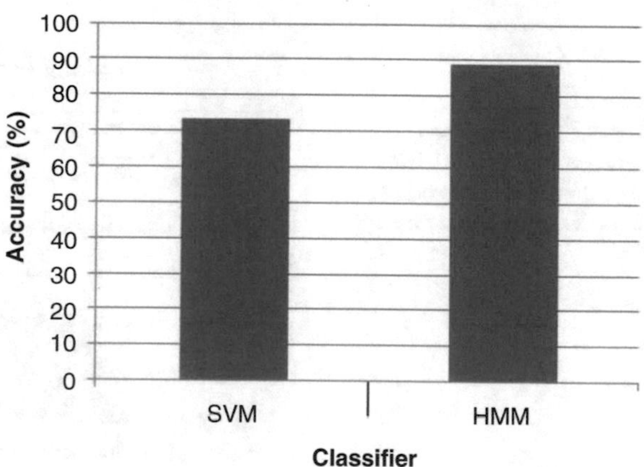

Fig. 5.11 Accuracy analysis between HMM and SVM for motor imagery tasks

in the dynamic nature of hemodynamic time series data, it gives the better result than SVM for all the dynamic tasks (finger tapping and motor imagery). From Fig. 5.12, we can conclude that error rate of HMM is less than SVM for both the tasks; thus HMM outperforms SVM in both the cases.

5.4.2 Comparative Analysis Between Ensemble Method Classifiers

5.4.2.1 Comparative Analysis Between Bagging, Boosting, and Voting Algorithm

In paper [11], the author has implemented a new voting algorithm and compared that new algorithm with bagging (Breiman's CART algorithm) and boosting (Quinlan C4.5 algorithm). In this paper, the author divides the error into two terms (bias

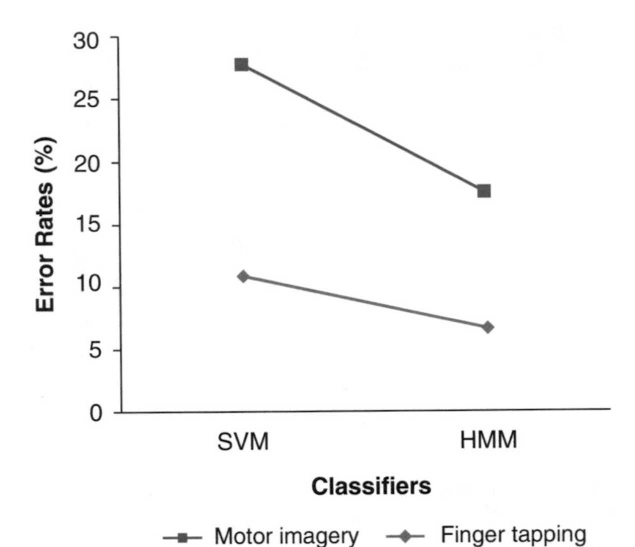

Fig. 5.12 Error rates analysis between HMM and SVM for finger tapping and motor imagery tasks

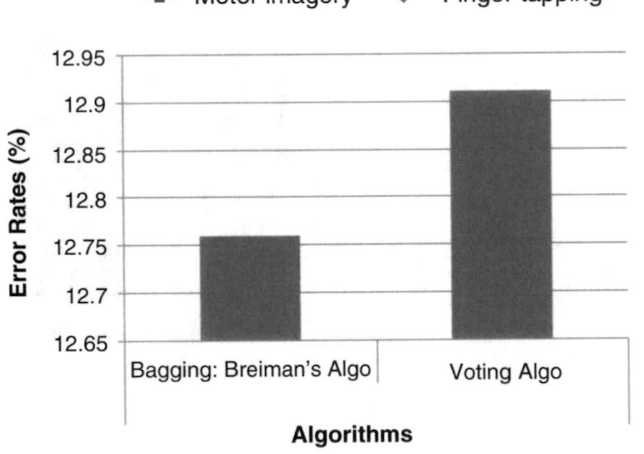

Fig. 5.13 Comparison between Breiman's (CART algorithm) [18] and voting algorithm [11] based on error rate

and variance) and tries to reduce that two terms based on some voting features (pruning the large tree, probabilistic estimate of the feature's coefficients, etc.). We have plotted the error rates and accuracy analysis between voting algorithm with boosting and bagging in Figs. 5.13 and 5.14.

From the figures (Figs. 5.13 and 5.14), we can conclude that voting algorithm returns higher accuracy than bagging (CART) and boosting (C4.5) algorithm by reducing the error rates and increasing the performance gain. After adding several variants into the voting algorithms, it is shown that voting algorithm performs better than bagging and boosting in stable and unstable methods. The author also compares the voting vs nonvoting methods based on the mean-squared error (MSE) and proved that voting method significantly reduces the MSE [11].

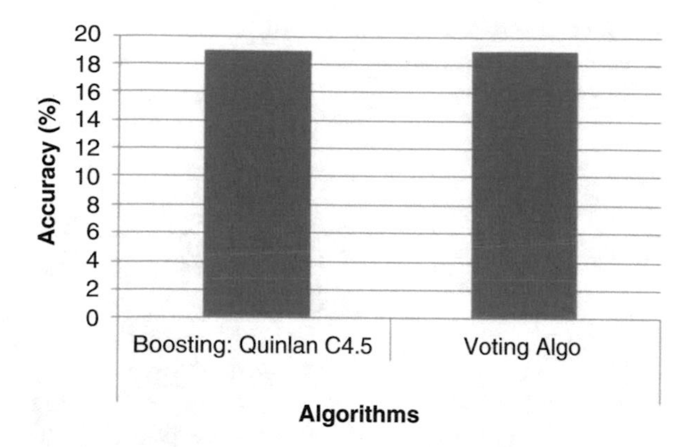

Fig. 5.14 Accuracy analysis between boosting (Quinlan C4.5) and voting algorithm

5.4.3 Comparative Analysis Between Basic and Ensemble Classifiers

5.4.3.1 Comparative Analysis Between SVM and Random Forest

In paper [12], the author has described Filter Bank Common Spatial Patterns (FBCSPs) with random forest to classify EEG motor imagery signal. This EEG activity is generated when a person imagines moving his hands. When a person tries to move his hands, then the μ and β frequency band of EEG signal activate and exhibit ERD and ERS movement depending on the increase or decrease of the amplitude value of μ or β bands. FBCSP algorithm has extracted the useful features (for imaging movement) after choosing proper frequency bands and then classifies those useful features using the classifier (random forest). Below we plot a comparison analysis between random forest and SVM in terms of accuracy and kappa value (used for accuracy measurement in multiple subjects).

From Fig. 5.15, we can conclude that random forest returns better classification result than SVM based on accuracy measure and kappa value. SVM can choose irrelevant features which are not useful in the classification process, whereas random forest (RF) avoids the feature selection steps as random forest is a combination of bagging and random selection of features, so random forest (for feature classification) along with FBCSP (for feature extraction) performs better than SVM. As SVM is sensitive to the irrelevant feature selection, error rates of SVM is more than random forest (Fig. 5.16).

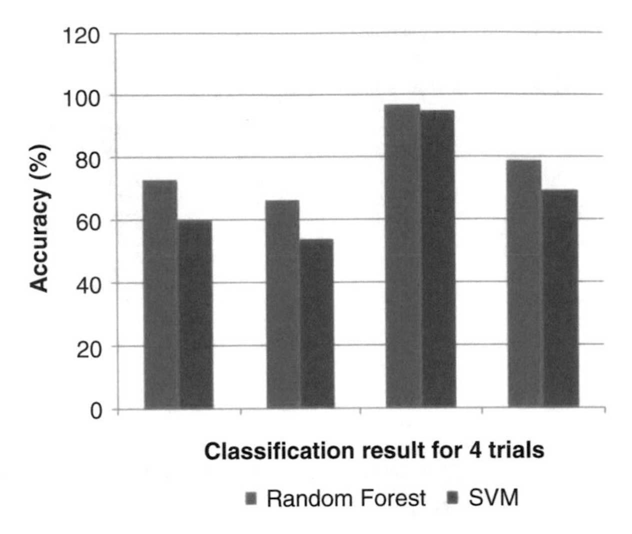

Fig. 5.15 Accuracy analysis between random forest and SVM with four trials

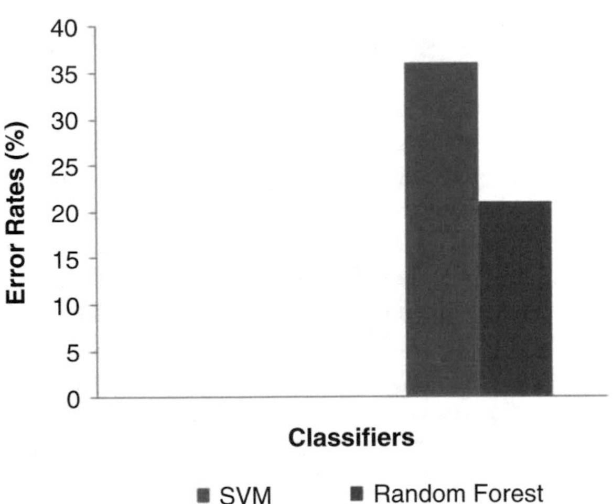

Fig. 5.16 Analysis based on error rates between SVM and random forest

5.4.4 Comparative Analysis Between Evolutionary Classifiers

5.4.4.1 Comparative Analysis Between ILS, HC, MA, and GA

In paper [3], the author has described different evolutionary algorithms (iterated local search (ILS), hill climber (HC), genetic algorithm (GA), memetic algorithm (MA)) with linkage information in features used in BCI. The linkage is a relationship between decision variables. This paper is highly emphasized on linkage distance between extracted features of all the algorithms. Here we have plotted a comparison analysis between all the evolutionary algorithms in terms of error rate.

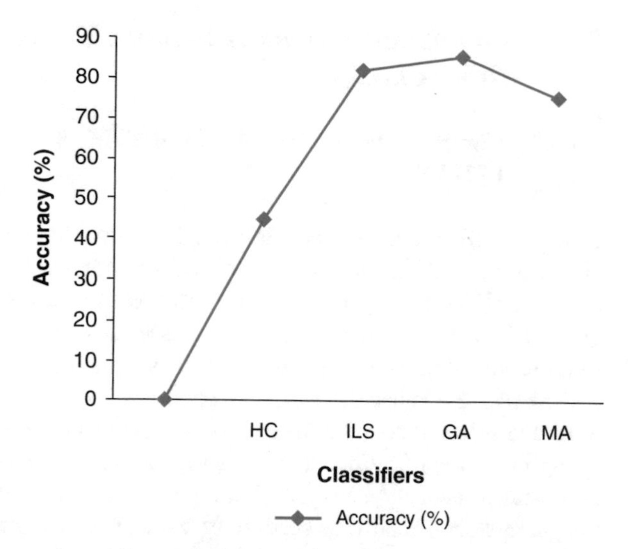

Fig. 5.17 Accuracy analysis between ILS, GA, HC, and MA for four trials

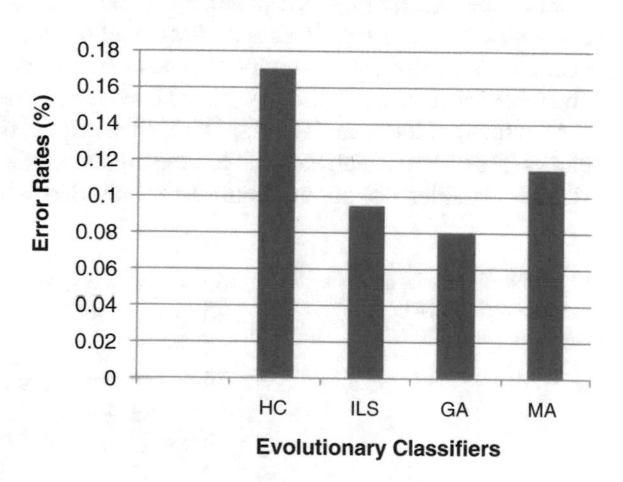

Fig. 5.18 Comparative analysis between HC, ILS, GA, and MA based on maximum error rate

From the figures (Figs. 5.17 and 5.18), we can conclude that genetic algorithm (GA) outperforms other three classifiers in terms of classification accuracy. After 30 runs of each algorithm, we found that GA is performing very well for linkage information with the minimum error rate, followed by ILS. GA returns highly fitted (minimum error rate in testing) linkage information of features using two steps (selection and crossover), so it always gives more classification accuracy than other classifiers.

5.4.5 *Comparative Analysis Between Basic, Ensemble, and Other Classifiers*

5.4.5.1 Comparative Analysis Between SVM, KNN, AdaBoost, and IT2FLS

In paper [73], the author has described the classification of motor imagery EEG signal using IT2FLS and wavelet transformation (WT). The paper is divided into two parts: (1) feature extraction using wavelet transformation and (2) classification using IT2FLS. In feature extraction phase, first, the raw EEG signal is divided into different channels. Then wavelet transformation is performed on each channel and selects the most discriminant coefficient of the wavelet using ROC curve. Then the selected coefficients are passed to the IT2FLS classifier. IT2FLS is used vastly in BCI applications due to its low computational cost. Here we perform the comparative analysis between IT2FLS with KNN, SVM, and AdaBoost based on the parameters (sensitivity, specificity, F-measure, accuracy).

From the figures (Figs. 5.19, 5.20, 5.21, and 5.22), we can conclude that IT2FLS classifier outperforms all other classifiers in all the cases except specificity analysis. IT2FLS has the highest sensitivity (true positive rate) than the other classifiers which gives more accuracy. IT2FLS has proved itself as the most stable classifier because it has the smallest interquartile range (IQR) than other classifiers. IT2FLS also has the highest F-measure which means it has the highest predictive power of classification. IT2FLS classifier also has minimum error rates due to highest sensitivity value.

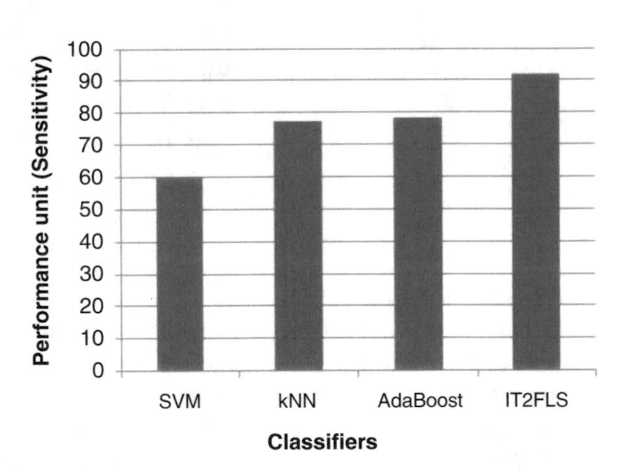

Fig. 5.19 Sensitivity analysis between SVM, KNN, AdaBoost, and IT2FLS

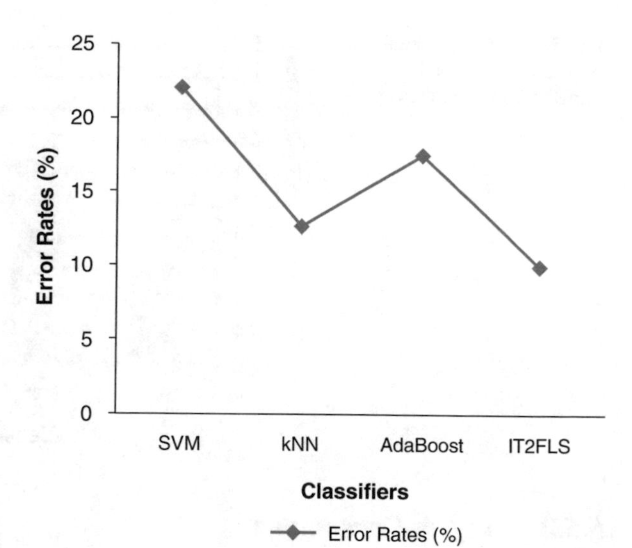

Fig. 5.20 Error analysis between SVM, KNN, AdaBoost, and IT2FLS

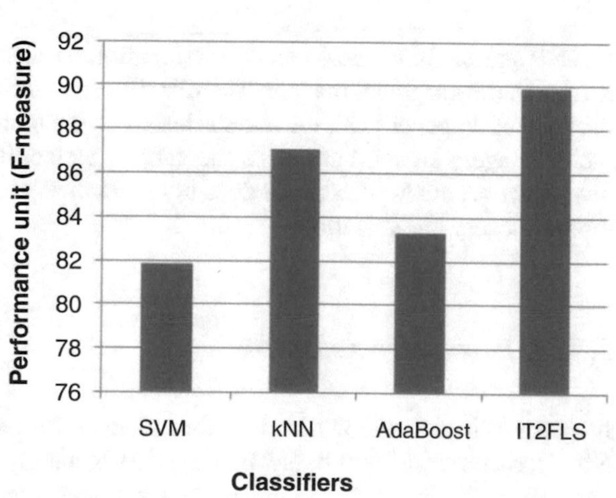

Fig. 5.21 F-measure analysis between SVM, KNN, AdaBoost, and IT2FLS

5.5 Open Problems

5.5.1 Selecting Proper Brain Signal

In paper [14], the author describes that changes in brain signal over time produce different patterns. ERD-based BCI could not identify some patterns which are recognized by SSVEP signal. Sometimes paralyzed people cannot produce a proper pattern using brain signals. So choosing appropriate brain signal is very important.

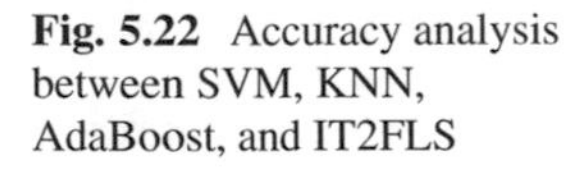
Fig. 5.22 Accuracy analysis between SVM, KNN, AdaBoost, and IT2FLS

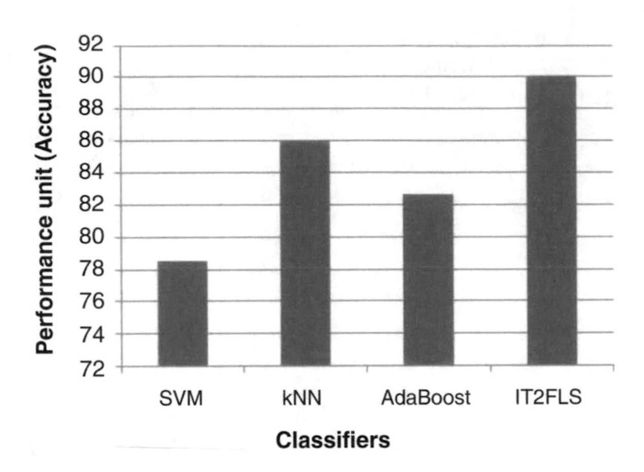

5.5.2 Subject Dependency

FBCSP and random forest-based classification do not give the best result for each feature in time segment manner. The classification accuracy is highly dependent on the subject. In paper [12], the author describes the binary classification of the EEG motor imagery signal, but due to the subject variability, this method has failed to give better accuracy. This paper does not contain any information about multi-class motor imagery classification.

5.5.3 Parameter Selection

In paper [9], the author has described that performance of FLDA, BLDA, and NN approaches in P300-based BCI system is mainly dependent on various input parameters like input data, number of electrodes, electrode configuration, feature vectors, etc. This paper illustrates the preprocessing steps (scaling of inputs, configuration of the electrodes, feature vector creation, etc.), which also affect the result of classification.

5.5.4 Small Training Set

If the training set is relatively small, then the training process cannot be completed and tested properly due to unavailability of samples. Genetic algorithm suffers from the high correlation of feature vectors over the small dataset, so the classification accuracy becomes poor [83].

5.5.5 Nonlinearity

The brain is a highly complex nonlinear system [1]. As the EEG signal is nonlinear and dynamic in nature, then the nonlinear classifier gets the advantage over the linear one. Feature extraction process is sometimes dependent on the dynamic behavior of the signal.

5.5.6 Noisy Environment

Noise is a big challenge to build a good BCI system. Nonstationary property of the brain signal sometimes causes noise. Noise can be originated either by repositioning the electrode positions or from the external events. In paper [29], the author has shown how noise impacts the performance of ensemble classifiers.

5.5.7 Hemodynamic Response

In paper [91], the author describes word speller application using NIRS signal. The signal is generated by rapid changes in the optical properties of cerebral tissue. The signal produces some hemodynamic response of different patterns for classification. The hemodynamic response is very slow to operate, and NIRS signal strength is highly affected by hair thickness and head movement of the user.

5.5.8 Feature Selection

Feature cluster taxonomy-based feature selection (FCTFS) method selects the suitable features from the different clusters to make the final subset. In high-dimensional dataset, this method cannot select the proper number of feature clusters. As a result feature redundancy and relevant features are not selected in the final subset. So the final classification process also does not return proper accuracy [41].

5.6 Conclusion

In this paper, we have completed a comprehensive survey of various techniques of feature extraction and classification methods used in BCI research. Many classification methods have been proposed and implemented in various brain-computer interfaces, and comparison of these methods for different BCI applications would be a useful task. This survey consists of the description of different classification methods (linear, nonlinear, ensemble-based) and how these classifiers are used in BCI design for feature selection, extraction, and classification purpose. We have done a comparative analysis among those classifiers used in BCI in terms of classification accuracy. We also figure out the open challenges with respect to this field and find out which classification methods have been used for a certain type of a BCI system. We hope that this survey helps the researchers about the extensive BCI research in the future.

Acknowledgements No research funding has been received for this survey work.

References

1. Abdulkader SN, Atia A, Mostafa MSM (2015) Brain computer interfacing: applications and challenges. Egyptian Inf J 16(2):213–230
2. Acharya UR, Sree SV, Alvin APC, Suri JS (2012) Use of principal component analysis for automatic classification of epileptic EEG activities in wavelet framework. Exp Syst Appl 39(10):9072–9078
3. Adair J, Brownlee A, Ochoa G (2017) Evolutionary Algorithms with Linkage Information for Feature Selection in Brain Computer Interfaces. In: Angelov P, Gegov A, Jayne C, Shen Q (eds) Advances in computational intelligence systems. Advances in intelligent systems and computing, vol 513. Springer, Cham
4. Ahangi A, Karamnejad M, Mohammadi N, Ebrahimpour R, Bagheri N (2013) Multiple classifier system for EEG signal classification with application to brain-computer interfaces. Neural Comput Appl 23(5):1319–1327
5. Akram F, Han SM, Kim TS (2015) An efficient word typing P300-BCI system using a modified T9 interface and random forest classifier. Comput Biol Med 56:30–36
6. Alamdari N, Haider A, Arefin R, Verma AK, Tavakolian K, Fazel-Rezai R (2016) A review of methods and applications of brain computer interface systems. In: 2016 IEEE International Conference on Electro Information Technology (EIT), May 2016. IEEE, pp 0345–0350
7. Al-Fahoum AS, Al-Fraihat AA (2014) Methods of EEG signal features extraction using linear analysis in frequency and time-frequency domains. In: ISRN Neuroscience 2014
8. Aydemir O (2016) Classification of 2-dimensional cursor movement imagery EEG signals. In: 2016 39th International Conference on Telecommunications and Signal Processing (TSP), June 2016. IEEE, pp 370–373
9. Bakhshi A, Ahmadifard A (2012) A comparison among classification accuracy of neural network, FLDA and BLDA in P 300-based BCI system. Int J Comput Appl 46(19):11–15
10. Bashar SK, Hassan AR, Bhuiyan MIH (2015) Identification of motor imagery movements from EEG signals using dual tree complex wavelet transform. In: 2015 International Conference on Advances in Computing, Communications and Informatics (ICACCI), Aug 2015. IEEE, pp 290–296
11. Bauer E, Kohavi R (1999) An empirical comparison of voting classification algorithms: bagging, boosting, and variants. Mach Learn 36(1):105–139
12. Bentlemsan M, Zemouri ET, Bouchaffra D, Yahya-Zoubir B, Ferroudji K (2014) Random forest and filter bank common spatial patterns for EEG-based motor imagery classification. In: 2014 5th International Conference on Intelligent Systems, Modelling and Simulation (ISMS), Jan 2014. IEEE, pp 235–238
13. Bhattacharyya S, Khasnobish A, Chatterjee S, Konar A, Tibarewala DN (2010) Performance analysis of LDA, QDA and KNN algorithms in left-right limb movement classification from

EEG data. In: 2010 International Conference on Systems in Medicine and Biology (ICSMB), Dec 2010. IEEE, pp 126–131
14. Bi L, Fan XA, Liu Y (2013) EEG-based brain-controlled mobile robots: a survey. IEEE Trans Hum-Mach Syst 43(2):161–176
15. Blankertz B, Lemm S, Treder M, Haufe S, Muller KR (2011) Single-trial analysis and classification of ERP components–a tutorial. NeuroImage 56(2):814–825
16. Bostanov V (2004) BCI competition 2003-data sets Ib and IIb: feature extraction from event-related brain potentials with the continuous wavelet transform and the t-value scalogram. IEEE Trans Biomed Eng 51(6):1057–1061
17. Boye AT, Kristiansen UQ, Billinger M, do Nascimento OF, Farina D (2008) Identification of movement-related cortical potentials with optimized spatial filtering and principal component analysis. Biomed Signal Process Control 3(4):300–304
18. Breiman L (1996) Bagging predictors. Mach Learn 24(2):123–140
19. Chakraborty S, Kumar S, Paul S, Kairi A (2017) A study of product trend analysis of review datasets using Naive Bayes, K-NN and SVM classifiers. Int J Adv Eng Manag 2(9):204–213. https://doi.org/10.24999/IJOAEM/02090047
20. Chen LL, Madhavan R, Rapoport BI, Anderson WS (2013) Real-time brain oscillation detection and phase-locked stimulation using autoregressive spectral estimation and time-series forward prediction. IEEE Trans Biomed Eng 60(3):753–762
21. Chiappa S, Barber D (2006) EEG classification using generative independent component analysis. Neurocomputing 69(7):769–777
22. Corralejo R, Hornero R, Alvarez D (2011) Feature selection using a genetic algorithm in a motor imagery-based brain computer interface. In: 2011 Annual International Conference of the IEEE on Engineering in Medicine and Biology Society, EMBC, Aug 2011. IEEE, pp 7703–7706
23. Cvetkovic D, Ubeyli ED, Cosic I (2008) Wavelet transform feature extraction from human PPG, ECG, and EEG signal responses to ELF PEMF exposures: a pilot study. Digit Signal Process 18(5):861–874
24. Darvishi S, Al-Ani A (2007) Brain-computer interface analysis using continuous wavelet transform and adaptive neuro-fuzzy classifier. In: 29th Annual International Conference of the IEEE Engineering in Medicine and Biology Society, Aug 2007. EMBS'07. IEEE, pp 3220–3223
25. Das Chakladar D, Chakraborty S (2017) Study and analysis of a fast moving cursor Control in a multithreaded way in brain computer interface. In: CICBA Conference, Kolkata
26. Demiralp T, Yordanova J, Kolev V, Ademoglu A, Devrim M, Samar VJ (1999) Time –frequency analysis of single-sweep event-related potentials by means of fast wavelet transform. Brain Lang 66(1):129–145
27. Devlaminck D, Wyns B, Boullart L, Santens P, Otte G (2009) Brain-computer interfaces: from theory to practice. In: ESANN
28. Dey L, Chakraborty S, Biswas A, Bose B, Tiwari S (2016) Sentiment analysis of review datasets using Naive Bayes' and K-NN classifier. Int J Inf Eng Electr Bus (IJIEEB) 8(4):54–62. https://doi.org/10.5815/ijieeb.2016.04.07
29. Dietterich TG (2000) An experimental comparison of three methods for constructing ensembles of decision trees: bagging, boosting, and randomization. Mach Learn 40(2):139–157
30. Duda RO, Hart PE, Stork DG (1995) Pattern classification and scene analysis, 2nd edn. Wiley Interscience, New York
31. Fatehi TAH, Suleiman ABR (2011) Features extraction techniques of EEG signals For BCI application, pp 35–40
32. Fatourechi M, Bashashati A, Ward RK, Birch GE (2005) A hybrid genetic algorithm approach for improving the performance of the LF-ASD brain computer interface. In: IEEE International Conference on Acoustics, Speech, and Signal Processing, Mar 2005. Proceedings (ICASSP'05), vol 5. IEEE, pp v-345

33. Fazel-Rezai R, Abhari K (2008) A comparison between a matrix-based and a region-based P300 speller paradigms for brain-computer interface. In: 30th Annual International Conference of the IEEE 2008 on Engineering in Medicine and Biology Society, Aug 2008. EMBS'08. IEEE, pp 1147–1150
34. Fazel-Rezai R, Abhari K (2009) A region-based P300 speller for brain-computer interface. Can J Electr Comput Eng 34(3):81–85
35. Friedman J, Hastie T, Tibshirani R (2001) The elements of statistical learning. Springer series in statistics, vol 1. Springer, New York, pp 241–249
36. Galar M, Fernandez A, Barrenechea E, Bustince H, Herrera F (2012) A review on ensembles for the class imbalance problem: bagging, boosting, and hybrid-based approaches. IEEE Trans Syst Man Cybern Part C (Appl Rev) 42(4):463–484
37. Gao JF, Yang Y, Lin P, Wang P, Zheng CX (2010) Automatic removal of eye-movement and blink artifacts from EEG signals. Brain Topogr 23(1):105–114
38. Garcia GN, Ebrahimi T, Vesin JM (2003) Support vector EEG classification in the Fourier and time-frequency correlation domains. In: First International IEEE EMBS Conference on Neural Engineering, Mar 2003. Conference Proceedings. IEEE, pp 591–594
39. Garrett D, Peterson DA, Anderson CW, Thaut MH (2003) Comparison of linear, nonlinear, and feature selection methods for EEG signal classification. IEEE Trans Neural Syst Rehabil Eng 11(2):141–144
40. Gavett S, Wygant Z, Amiri S, Fazel-Rezai R (2012) Reducing human error in P300 speller paradigm for brain-computer interface. In: 2012 Annual International Conference of the IEEE on Engineering in Medicine and Biology Society (EMBC), Aug 2012. IEEE, pp 2869–2872
41. Goswami S, Das AK, Chakrabarti A, Chakraborty B (2017) A feature cluster taxonomy based feature selection technique. Exp Syst Appl 79:76–89
42. Grosse-Wentrup M, Buss M (2008) Multiclass common spatial patterns and information theoretic feature extraction. IEEE Trans Biomed Eng 55(8):1991–2000
43. Hazrati MK, Erfanian A (2010) An online EEG-based brain-computer interfaces for controlling hand grasp using an adaptive probabilistic neural network. Med Eng Phys 32(7):730–739
44. Herman P, Prasad G, McGinnity TM (2006) Investigation of the type-2 fuzzy logic approach to classification in an EEG-based brain-computer interface. In: 27th Annual International Conference of the Engineering in Medicine and Biology Society, Jan 2005. IEEE-EMBS'05. IEEE, pp 5354–5357
45. Hoffmann U, Garcia G, Vesin JM, Diserens K, Ebrahimi T (2005) A boosting approach to P300 detection with application to brain-computer interfaces. In: 2nd International IEEE EMBS Conference on Neural Engineering, Mar 2005. Conference Proceedings. IEEE, pp 97–100
46. Hsu WY (2010) EEG-based motor imagery classification using neuro-fuzzy prediction and wavelet fractal features. J Neurosci Methods 189(2):295–302
47. Hsu WY, Lin CC, Ju MS, Sun YN (2007) Wavelet-based fractal features with active segment selection: application to single-trial EEG data. J Neurosci Methods 163(1):145–160
48. Huan NJ, Palaniappan R (2004) Neural network classification of autoregressive features from electroencephalogram signals for brain–computer interface design. J Neural Eng 1(3):142
49. Ishfaque A, Awan AJ, Rashid N, Iqbal J (2013) Evaluation of ANN, LDA and decision trees for EEG based brain computer interface. In: 2013 IEEE 9th International Conference on Emerging Technologies (ICET), Dec 2013. IEEE, pp 1–6
50. James G, Witten D, Hastie T, Tibshirani R (2013) An introduction to statistical learning, vol 112. Springer, New York
51. Kaur M, Ahmed P, Rafiq MQ (2012) Technology development for unblessed people using BCI: a survey. Int J Comput Appl 40(1):18–24
52. Kayikcioglu T, Aydemir O (2010) A polynomial fitting and k-NN based approach for improving classification of motor imagery BCI data. Pattern Recogn Lett 31(11):1207–1215
53. Khosrowabadi R, Quek C, Ang KK, Tung SW, Heijnen M (2011) A brain-computer interface for classifying EEG correlates of chronic mental stress. In: The 2011 International Joint Conference on Neural Networks (IJCNN), July 2011. IEEE, pp 757–762

54. Kolodziej M, Majkowski A, Rak R (2011) A new method of EEG classification for BCI with feature extraction based on higher order statistics of wavelet components and selection with genetic algorithms. Adaptive and Natural Computing Algorithms, pp 280–289
55. Kottaimalai R, Rajasekaran MP, Selvam V, Kannapiran B (2013) EEG signal classification using principal component analysis with neural network in brain computer interface applications. In: 2013 International Conference on Emerging Trends in Computing, Communication and Nanotechnology (ICE-CCN), Mar 2013. IEEE, pp 227–231
56. Krusienski DJ, McFarland DJ, Wolpaw JR (2006) An evaluation of autoregressive spectral estimation model order for brain-computer interface applications. In: 28th Annual International Conference of the IEEE on Engineering in Medicine and Biology Society, Aug 2006. EMBS'06. IEEE, pp 1323–1326
57. Lakshmi MR, Prasad DT, Prakash DVC (2014) Survey on EEG signal processing methods. Int J Adv Res Comput Sci Softw Eng 4(1):84–91
58. Lawhern V, Hairston WD, McDowell K, Westerfield M, Robbins K (2012) Detection and classification of subject-generated artifacts in EEG signals using autoregressive models. J Neurosci Methods 208(2):181–189
59. Lee H, Choi S (2003) PCA+ HMM+ SVM for EEG pattern classification. In: Seventh International Symposium on Signal Processing and Its Applications, July 2003. Proceedings, vol 1. IEEE, pp 541–544
60. Li Y, Ma Z, Lu W, Li Y (2006) Automatic removal of the eye blink artifact from EEG using an ICA-based template matching approach. Physiol Meas 27(4):425
61. Li Y, Guan C, Li H, Chin Z (2008) A self-training semi-supervised SVM algorithm and its application in an EEG-based brain computer interface speller system. Pattern Recogn Lett 29(9):1285–1294
62. Li K, Sankar R, Arbel Y, Donchin E (2009) Single trial independent component analysis for P300 BCI system. In: Annual International Conference of the IEEE Engineering in Medicine and Biology Society, Sept 2009. EMBC'09. IEEE, pp 4035–4038
63. Lin CJ, Hsieh MH (2009) Classification of mental task from EEG data using neural networks based on particle swarm optimization. Neurocomputing 72(4):1121–1130
64. Lotte F, Congedo M, Lecuyer A, Lamarche F, Arnaldi B (2007) A review of classification algorithms for EEG-based brain-computer interfaces. J Neural Eng 4(2):R1
65. McFarland DJ, Krusienski DJ, Wolpaw JR (2006) Brain-computer interface signal processing at the Wadsworth center: mu and sensorimotor beta rhythms. Prog Brain Res 159:411–419
66. McFarland DJ, Anderson CW, Muller KR, Schlogl A, Krusienski DJ (2006) BCI meeting 2005-workshop on BCI signal processing: feature extraction and translation. IEEE Trans Neural Syst Rehabil Eng 14(2):135–138
67. Mousavi EA, Maller JJ, Fitzgerald PB, Lithgow BJ (2011) Wavelet common spatial pattern in asynchronous offline brain computer interfaces. Biomed Signal Process Control 6(2):121–128
68. Mousavi M, Koerner AS, Zhang Q, Noh E, de Sa VR (2017) Improving motor imagery BCI with user response to feedback. Brain-Comput Interfaces 4(1–2):74–86
69. Muhl C, Allison B, Nijholt A, Chanel G (2014) A survey of affective brain computer interfaces: principles, state-of-the-art, and challenges. Brain-Comput Interfaces 1(2):66–84
70. Muller KR, Anderson CW, Birch GE (2003) Linear and nonlinear methods for brain-computer interfaces. IEEE Trans Neural Syst Rehabil Eng 11(2):165–169
71. Mustafa M, RTaid MN, Murat ZH, Sulaiman N (2012) Comparison between KNN and ANN classification in brain balancing application via spectrogram image. JCSCM 2(4):17–22
72. Nasehi S, Pourghassem H (2013) Mental task classification based on HMM and BPNN. In: 2013 International Conference on Communication Systems and Network Technologies (CSNT), Apr 2013. IEEE, pp 210–214
73. Nguyen T, Khosravi A, Creighton D, Nahavandi S (2015) EEG signal classification for BCI applications by wavelets and interval type-2 fuzzy logic systems. Exp Syst Appl 42(9):4370–4380
74. Nicolas-Alonso LF, Gomez-Gil J (2012) Brain computer interfaces, a review. Sensors 12(2):1211–1279

75. Obermaier B, Guger C, Neuper C, Pfurtscheller G (2001) Hidden Markov models for online classification of single trial EEG data. Pattern Recogn Lett 22(12):1299–1309
76. Ofner P, Müller-Putz GR, Neuper C, Brunner C (2011) Comparison of feature extraction methods for brain-computer interfaces. NA. https://pdfs.semanticscholar.org/0b1c/ddbd83820d7f30a82ca340cf027c3cb17f87.pdf
77. Peterson DA, Knight JN, Kirby MJ, Anderson CW, Thaut MH (2005) Feature selection and blind source separation in an EEG-based brain-computer interface. EURASIP J Adv Signal Process 2005(19):218613
78. Pires G, Castelo-Branco M, Nunes U (2008) Visual P300-based BCI to steer a wheelchair: a Bayesian approach. In: 30th Annual International Conference of the IEEE on Engineering in Medicine and Biology Society, Aug 2008. EMBS'08. IEEE, pp 658–661
79. Platt JC, Nello Cristianini N, Shawe-Taylor J (2000) Large margin DAG for multiclass classification. In: Solla SA, Leen TK, Muller K-R (eds) Advances in neural information processing systems 12. MIT Press, Cambridge, MA, pp 547–553
80. Rak RJ, Kolodziej M, Majkowski A (2012) Brain-computer interface as measurement and control system the review paper. Metrol Meas Syst 19(3):427–444
81. Rakotomamonjy A, Guigue V (2008) BCI competition III: dataset II-ensemble of SVMs for BCI P300 speller. IEEE Trans Biomed Eng 55(3):1147–1154
82. Ramadan RA, Refat S, Elshahed MA, Ali RA (2015) Basics of brain computer interface. In: Hassanien A, Azar A (eds) Brain-computer interfaces. Intelligent systems reference library, vol 74. Springer, Cham
83. Rejer I (2015) Genetic algorithm with aggressive mutation for feature selection in BCI feature space. Pattern Anal Appl 18(3):485–492
84. Roman-Gonzalez A (2012) EEG signal processing for BCI applications. In: Hippe ZS, Kulikowski JL, Mroczek T (eds) Human – computer systems interaction: backgrounds and applications 2. Advances in intelligent and soft computing, vol 98. Springer, Berlin/Heidelberg.
85. Samek W, Vidaurre C, Muller KR, Kawanabe M (2012) Stationary common spatial patterns for brain–computer interfacing. J Neural Eng 9(2):026013
86. Sanei S, Chambers JA (2013) EEG signal processing. Wiley, Somerset
87. Scherer R, Muller GR, Neuper C, Graimann B, Pfurtscheller G (2004) An asynchronously controlled EEG-based virtual keyboard: improvement of the spelling rate. IEEE Trans Biomed Eng 51(6):979–984
88. Schlogl A, Lee F, Bischof H, Pfurtscheller G (2005) Characterization of four-class motor imagery EEG data for the BCI-competition 2005. J Neural Eng 2(4):L14
89. Schwenk H, Bengio Y (2000) Boosting neural networks. Neural Comput 12(8):1869–1887
90. Shi X, Xu G, Shen F, Zhao J (2015) Solving the data imbalance problem of P300 detection via random under-sampling bagging SVMs. In: 2015 International Joint Conference on Neural Networks (IJCNN), July 2015. IEEE, pp 1–5
91. Sitaram R, Zhang H, Guan C, Thulasidas M, Hoshi Y, Ishikawa A, Birbaumer N (2007) Temporal classification of multichannel near-infrared spectroscopy signals of motor imagery for developing a brain-computer interface. NeuroImage 34(4):1416–1427
92. Suleiman ABR, Fatehi TAH (2007) Features extraction techniques of EEG signal for BCI applications. Faculty of Computer and Information Engineering Department College of Electronics Engineering, University of Mosul
93. Sykacek P, Roberts SJ, Stokes M (2004) Adaptive BCI based on variational Bayesian Kalman filtering: an empirical evaluation. IEEE Trans Biomed Eng 51(5):719–727
94. Thomas E, Fruitet J, Clerc M (2013) Combining ERD and ERS features to create a system-paced BCI. J Neurosci Methods 216(2):96–103
95. Thulasidas M, Guan C, Wu J (2006) Robust classification of EEG signal for brain-computer interface. IEEE Trans Neural Syst Rehabil Eng 14(1):24–29
96. Ting W, Guo-zheng Y, Bang-hua Y, Hong S (2008) EEG feature extraction based on wavelet packet decomposition for brain computer interface. Measurement 41(6):618–625

97. Van Gerven M, Farquhar J, Schaefer R, Vlek R, Geuze J, Nijholt A, Desain P (2009) The brain-computer interfaces cycle. J Neural Eng 6(4):041001
98. Vidaurre C, Kramer N, Blankertz B, Schlogl A (2009) Time domain parameters as a feature for EEG-based brain-computer interfaces. Neural Netw 22(9):1313–1319
99. Von Bunau P, Meinecke FC, Kiraly FC, Muller KR (2009) Finding stationary subspaces in multivariate time series. Phys Rev Lett 103(21):214101
100. Wang J, Xu G, Wang L, Zhang H (2010) Feature extraction of brain-computer interface based on improved multivariate adaptive autoregressive models. In: 2010 3rd International Conference on Biomedical Engineering and Informatics (BMEI), Oct 2010, vol 2. IEEE, pp 895–898
101. Wolpaw J, Wolpaw EW (eds) (2012) Brain-computer interfaces: principles and practice. OUP, Oxford/New York
102. Xing S, McCardle R, Xie S (2012) Reading the mind: the potential of electroencephalography in brain computer interfaces. In: 2012 19th International Conference on Mechatronics and Machine Vision in Practice (M2VIP), Nov 2012. IEEE, pp 275–280
103. Yang R, Gray DA, Ng BW, He M (2009) Comparative analysis of signal processing in brain computer interface. In: 4th IEEE Conference on Industrial Electronics and Applications, May 2009, ICIEA'09. IEEE, pp 580–585
104. Yang J, Singh H, Hines EL, Schlaghecken F, Iliescu DD, Leeson MS, Stocks NG (2012) Channel selection and classification of electroencephalogram signals: an artificial neural network and genetic algorithm-based approach. Artif Intell Med 55(2):117–126
105. Yuan H, He B (2014) Brain-computer interfaces using sensorimotor rhythms: current state and future perspectives. IEEE Trans Biomed Eng 61(5):1425–1435
106. Zhang H, Guan C, Wang C (2008) Asynchronous P300-based brain–computer interfaces: a computational approach with statistical models. IEEE Trans Biomed Eng 55(6):1754–1763

Chapter 6
A Multilingual Framework for a Question Answering System for the Support of Ticket Data: An Extended Approach

Suyog Trivedi, Balaji Jagan, and Gopichand Agnihotram

6.1 Introduction

Semantic matching is one of the important tasks in many natural language processing (NLP) applications, such as information retrieval [1], question answering [2], etc. Considering question answering system as an example, given a pair of question and answer, a matching function is required to determine the degree of matching between these two sentences. Moreover, matching the user question with the list of question-answer pairs is a difficult task where the user queries are not always complete and grammatically and syntactically correct. In addition, capturing the intent of the query involves finding the semantic information at a deeper level. Nowadays, deep neural network-based models have been applied to overcome such issues. A lot of deep learning models represent the question and answer in a single distributed representation and then compute similarities between the query vector and the question-answer pair vectors to output the matching score.

In order to properly represent words in a continuous space, the idea of a neural model [3] is employed to enable jointly learn embedding of words into an n-dimensional vector space and to use these vectors to predict how likely a word is given its context. Skip-gram model [4] is a widely used approach to compute such an embedding. When skip-gram networks are optimized via gradient ascent, the derivatives modify the word embedding matrix L R (nV), where V is the size of the vocabulary. The word vectors inside the embedding matrix capture distributional syntactic and semantic information via the word co-occurrence statistics [3, 4]. Once this matrix is learned on an unlabeled corpus, it can be used for subsequent tasks by using each words vector ($\underline{v}$, $\underline{w}$) to represent the corresponding word.

S. Trivedi (✉) · B. Jagan · G. Agnihotram
Wipro Technologies, India
e-mail: suyog.trivedi1@wipro.com; balaji.jagan@wipro.com; gopichand.agnihotram@wipro.com

X. Li, K.-C. Wong (eds.), *Natural Computing for Unsupervised Learning*,
Unsupervised and Semi-Supervised Learning,
https://doi.org/10.1007/978-3-319-98566-4_6

In this chapter, we describe a multilingual framework on semantic matcher based on the neural embedding models of different languages. An approach on semantic matcher for English language is adopted and further enhanced to support other languages [5]. The framework is broadly categorized into two parts: language dependent and language independent. The tasks/modules that heavily depend on the language-specific properties/grammars, natural language processing components, fall under language-dependent part. In particular, NLP tasks, such as stemmer, POS tagger, lemmatizer, sentence parser, etc., are language dependent. The tasks that fall under language-independent part are indexing, search, and ranking which purely involves statistical and mathematical models. The unlabeled corpus is preprocessed with the basic NLP tasks such as POS tagging and lemmatization using separate language-specific components. The random word vectors are built for both the questions and answers on top of the tagged data. The neural word embeddings for each language-specific data are created with the help of the pre-trained spaCy models.

In the same way, the query vectors are created and matched with the random vectors of the reference corpus. The cosine similarity measure is then applied on the matched vectors to compute the similarity and perform the ranking. The matched items with low ranking are filtered out and the remaining top K results are displayed.

The rest of the chapter is organized as follows. Section 6.2 describes the literature survey on the recent works carried out on semantic matching. Section 6.3 discuss the proposed framework on how semantic matching is carried out using the distributed vector representation for various languages. Section 6.3 also deals with the distributed semantic models which discusses about building the word vectors using the pre-trained vector-based neural models and GloVe word vector representation and describes the semantic similarity matching of query and the questions. Section 6.4 discusses the performance of the proposed approach in evaluation, and Sect. 6.5 concludes the advantages of the framework with future enhancements.

6.2 Literature Review

In this section, we discuss the recent works carried out on semantic mapping of query with document, ads, passages, etc. using distributed vector representation.

A neural network-based approach has been implemented with translation model and word embedding model to capture the word-to-word relation called a Word Embedding Correlation (WEC) model. The words in the query are mapped to vectors to identify the co-occurrence relationships between words. The word-level relationship is extended to the sentence level to calculate the relevance between question and answers [6]. An unsupervised approach is proposed for identifying the senses of search queries and performs the semantic clustering on each search engine page result. The word sense disambiguation is performed with the use of distributed word vector representations with the help of prediction-based neural embedding models [7]. A bidirectional LSTM has been applied to tackle the problem of matching two sentences with multiple positional sentence representations in which the contextual information is captured. The interaction between two representations is performed through k-max pooling and a multilayer perceptron by which the matching score is generated. Using this positional independent matching procedure, any part of the queries can be matched. This type of approach is useful in handling the data of fully and/or partially free-word order languages [8].

A semantic embedding-based approach has been proposed for extracting queries and ads in which the embedding is learned using various components such as search queries, clicked ads, search links, dwell time, and implicit negative signals. In case of the absence of information on new ads, the vector learns the context information from the textual content of the new ads including the bid term context vector [9].

A novel semantic matching-based retrieval model is proposed based on the Bag-of-Words Embedding (BoWE) representation. The semantic matching between queries and documents can be viewed as a nonlinear word transportation (NWT) problem (based on document word capacity and transportation profit) [10].

A semantic matching approach has been introduced which uses skip gram model, in which a distributed representation learned with neural networks for matching questions and answers based on their semantic similarity. The question and answers are represented at different linguistic levels to extract various features that overlap each other. The overlapping features are then used to obtain the linguistic similarity [11].

An approach on semantic mapping has been proposed to translate a natural language question into a SQL queries where the semantic mapping is carried out with the use of syntactic analyzer. The syntactic trees of questions and queries are represented as relational pairs and are encoded using the SVM-based kernel functions [12].

6.3 Semantic Matcher: A Multilingual Framework

This section deals with the semantic matching of the user utterances and produces top K questions for a search query. The semantic matching of a query is determined using the word embeddings with the help of pre-trained models such as spaCy and GloVe vectors. These models are used to detect the closest match of the user search query. The method proposes an idea of building word-based vector representation of the given query. The words and its context information are captured through these vectors where the similarity between the words is measured through the vector comparison. In addition, words that are semantically similar and/or semantically related to the existing query-words but are not part of the vectors are added to the word embedding vectors incrementally. Two words are considered to be semantically similar and/or semantically related if their distributional vector representation are similar based on some established col- location measure. The semantic similarity between the query word vectors and the document word vectors is computed using a traditional measure: cosine similarity. The cosine distance between the trained-model word vectors and the query word vectors is used as a feature to identify best possible matches for the given query. The detailed diagram of semantic matcher framework is shown in Fig. 6.1.

The semantic matcher framework shown in Fig. 6.1 describes the overall flow in which the language detector is used to automatically detect the language and to load language based NLP components which are used for preprocessing. With this information, it further proceeds to load the multi-language models to build the random vectors. Once the random vectors are created with the help of spaCy and GloVe word vector representation, all the extracted information is indexed. In this framework, we use common indexing mechanism for all language data. While building common index for different languages, it is important to add the language identifier as an additional field to avoid the complexities during search. Thus, a language identifier is added to each entry in the index. While performing the search using the user query, the language identifier is also passed as a parameter to search in the index which makes the search easy to obtain the appropriate results. In this framework, we start with an separate corpora for each language which can be defined as below.

The corpus (C_L) of different languages is constructed using the questions and answers and is defined as

$$C_L = (Q_1, A_1), (Q_2, A_2), \ldots (Q_n, A_n)_L \tag{6.1}$$

where Q_i is the ith question and i = 1,2,. . . n and A_i is the ith answer and i = 1,2,. . . n.

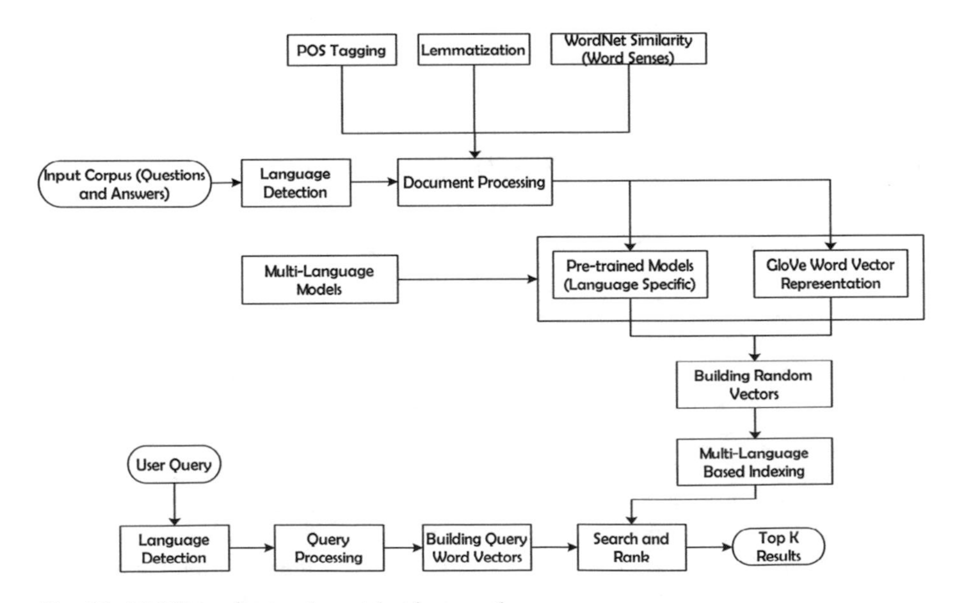

Fig. 6.1 Multilingual semantic matcher framework

6.3.1 Preprocessing

The data under consideration provides a list of questions and answers. The preprocessing module consists of tokenizer, lemmatizer, and POS tagger for each language under consideration. A language detector is also used to automatically detect the language of the input data and the query, respectively. This is useful in loading the corresponding language-specific models. The tokenizer tokenizes all the words in the question and answers.

For instance, the nltk WordNet Lemmatizer is used to obtain the lemma of each word with the help of the nltk POS tagger for English words. If the word is not in the lemmatizer, then the actual word itself is returned as a lemma. In order to obtain the correct lemma of a word in a sentence, the POS tag of a word is also considered while finding out the lemma through WordNet lemmatizer. Then these words are given to the spaCy pre-trained model to create the word vectors where the semantic matching criterion is achieved along with the help of the GloVe vector representation and semantic similarity measure. Similarly, the language-specific components are used, respectively, for performing the basic NLP preprocessing tasks of the corresponding languages. The question-answer vector (for each language) represented after preprocessing is as given in (6.2) below:

$$(Q_{\mathrm{i}}, A_{\mathrm{i}})_{\mathrm{L}} = (S_{\mathrm{q}}, S_{\mathrm{a}})_{\mathrm{L}} = (w_{1\mathrm{q}}, w_{2\mathrm{q}}), \ldots (w_{\mathrm{nq}}), (w_{1\mathrm{a}}, w_{2\mathrm{a}}), \ldots (w_{\mathrm{na}})_{\mathrm{L}} \tag{6.2}$$

where S_{q} is the sentence corresponding to question and S_{a} is the sentence corresponding to answer. w_{iq} is the ith word corresponding to the question after preprocessing and similarly w_{ia} is the ith word corresponding to the answer, i = 1, 2, ...

After preprocessing these words are given to the spaCy pre-trained model to create the word embeddings/feature vectors where the semantic matching criterion is accomplished along with the help of the GloVe vector representation and semantic similarity measure. The next step/subsection discusses the word embedding/feature vector creation using the corpus after preprocessing.

6.3.2 Word Embedding/Feature Vector Creation

In this method, we have adopted the pre-trained neural embedding model for building the random word vectors. Word vectors encode semantic meaning and capture many different degrees of similarity [13]. There are a variety of computational models that implement the distributional hypothesis, including word2vec, GloVe, dependency-based word embeddings, and random indexing.

In this section, we use the dependency-based word embeddings implemented in spaCy for Troubleshooting data that consists of a set of questions and answers. This set is used as a training data to build the spaCy word embedding model with the help of the already built pre-trained model. The pre-trained model is available in spaCy where the model uses huge corpus comprised of news articles, Wikipedia documents, weblogs, newsgroups, blogs, etc. for building the model. The spaCy model has different features such POS tagger, NER (named-entity recognition), and sentiment analysis, along with creating semantic word embeddings. The spaCy model builds semantic word embeddings/feature vectors and internally uses the GloVe vectors and computes the top list words matching with distance measures such as cosine similarity and Euclidian distance approach. The large pre-trained spaCy model provides word vectors for about a million words in English language.

Along with the use of spaCy model, the question-answer set is used to build the domain-specific word vector representation using the one million word vectors. To build the word vector representation, the sentences in the corpus are preprocessed to derive the words and trained with the Glove framework to build word-vectors for domain-specific words.

The word embeddings are initialized to the 300-dimensional GloVe pre-trained vectors [14] on the Wikipedia 2014 and Gigaword 5 datasets. The weights for the cells are initialized using Xavier initialization, while the biases are zero-initialized.

Once the word vectors are created from the data, the similar procedure is applied to the query to build the query word vectors. The query word vectors are then matched with the random vectors in which the semantic similarity is computed and the closest match of results are obtained.

6.3.3 GloVe Word Vector Representation

GloVe is an unsupervised machine learning algorithm for obtaining vector representa- tions for words. The GloVe builds the training model using the aggregated global word-word co-occurrence statistics from a corpus. The method uses semantic similarity words using Euclidian distance or cosine similarity measure to build the resulting word representation with the word vector space.

In the GloVe vector representation for question answering model uses question answering corpus and builds the co-occurrence of unigram, bigram, and trigram words using the GloVe to compute the words of 300-dimensional feature vectors. Here we use the GloVe to train for the words of our corpus which are not found in spaCy model.

$$(Q_{\mathrm{i}}, A_{\mathrm{i}})_{\mathrm{L}} = (w_{1\mathrm{q}^1}, w_{2\mathrm{q}^1}), \ldots (w_{\mathrm{nq}^1}), (w_{1\mathrm{a}^1}, w_{2\mathrm{a}^1}), \ldots (w_{\mathrm{na}^1})_{\mathrm{L}} \tag{6.3}$$

$$(Q_i, A_i)_L = (S_q, S_a)_L = (w_{1q}, w_{2q}), \ldots (w_{nq}), (w_{1a}, w_{2a}), \ldots (w_{na})_L \bigcup$$
$$(w_{1q^1}, w_{2q^1}), \ldots (w_{nq^1}), (w_{1a^1}, w_{2a^1}), \ldots (w_{na^1})_L \quad (6.4)$$

where $(w_{1q^1}, w_{2q^1}), \ldots (w_{nq^1})$ are the words form question which are not part of spaCy model. $(w_{1a^1}, w_{2a^1}), \ldots (w_{na^1})$ are the words from answers which are not part of spaCy model.

$$w_{1q^1} = [0.0643, 0.5257, \ldots, 0.3679]_{1\times 300} \quad (6.5)$$

Similarly,

$$w_{1a^1} = [0.167175, 0.2875, \ldots 0.16753]_{1\times 300} \quad (6.6)$$

$$w_{1q^1} = [0.299463, 0.317821, \ldots 0.26345]_{1\times 300} \quad (6.7)$$

The feature vectors of the question answering corpus will be created using spaCy and Glove vector as explained in the above steps (Sects. 6.3.1, 6.3.2, and 6.3.3). The next subsection discusses the user query processing in above steps or Sects. 6.3.1, 6.3.2, and 6.3.3.

6.3.4 Indexing

In order to obtain the precise retrieval of questions (both exact and partial match), indexing of questions and answers is necessary to make the retrieval process more efficient. Indexing is an intermediate representation between query and the document base. Indexing in a multilingual environment can be divided in three steps: (6.1) language-dependent canonical term extraction (including stop-words elimination, stemming and/or lemmatization, word-sense disambiguation), (6.2) finding language- neutral term, and (6.3) update of the term-document association lattice. In the pro- posed framework, the first step is performed with the help of language-specific com- ponents for each language. The language identifier is used to detect the language of the input data and user query as well. This language identifier is considered along with the metadata information while indexing the questions and answers. Finally, the term-document association is updated incrementally while performing indexing. During creation of multilingual index, each term is indexed with various fields as part of the metadata. Some of the fields include terms, POS tags, entity tags, confidence scores, frequency scores, language identifier, etc. All of these metadata information are important while performing the search and rank.

6.3.5 Query Processing

The user queries are processed through the above steps, and the query word vectors are created, respectively. With the help of language detector, the language of the user queries is automatically detected, and the corresponding language models are enabled for building the query word vectors. For instance, the user query U_t consists of a set of sentence sequences or a unique sentence. The user query is first passed to the language detector where the language is identified automatically. Based on the language detected, the corresponding language-specific natural language components are loaded to remove stopwords, label the POS tags, etc.

$$U_t = (S_1, S_2, \ldots S_n) = (w_1, w_2, \ldots w_n) \tag{6.8}$$

Here $(S_1, S_2, \ldots, S_n)$ are the sentences from the utterance, and it can be unique sentence and sequence of sentence. $(w_1, w_2, \ldots, w_n)$ are the words extracted after preprocessing step. The spaCy feature vector of words in the utterance is given as

$$w_{iS} = [0.356557, 0.348621, \ldots 0.569231]_{1\times 300} \tag{6.9}$$

$$w_{nS} = [0.002453, 0.026352, \ldots 0.0369124]_{1\times 300} \tag{6.10}$$

Here, $(w_{1S}, w_{2S}, \ldots, w_{nS})$ are the feature vectors of the word representation using spaCy model. There may be few words wherein spaCy model is not able to address those words that will be addressed using GloVe feature vector as explained in Sect. 6.3.3. The GloVe feature vectors are shown below.

$$w_{iG} = [0.378294, 0.629482, \ldots 0.72359]_{1\times 300} \tag{6.11}$$

$$w_{nG} = [0.452625, 0.252418, \ldots 0.001352]_{1\times 300} \tag{6.12}$$

Here, $(w_{1G}, w_{2G}, \ldots, w_{nG})$ are the words that represent GloVe feature vectors. The utterance U_t is rewritten as given below

$$U_t = (w_1, w_2, \ldots w_n) = (w_{1S}, w_{2S}, \ldots w_{nS}) \bigcup (w_{1G}, w_{2G}, \ldots w_{nG}) \tag{6.13}$$

Once the user query/utterance is encoded into a feature vector representation, it is matched with the question answering corpus feature vectors created in Sects. 6.3.1, 6.3.2, and 6.3.3 using the pre-trained models of spaCy and GloVe vector representation. The next subsection discusses the matching of the feature vectors.

6.3.6 Semantic Similarity Matching

The framework makes use of cosine similarity or Euclidian distance methods to measure the distance between the feature vectors of user query/question/utterance and question answering corpus. Here we adapt cosine similarity measure to compute the distance, and cosine similarity is as given below:

$$(\underline{a}, \underline{b}) = \|a\| \|b\| \cos(\theta) \tag{6.14}$$

$$\cos(\theta) = \frac{a \cdot b}{\|a\| \|b\|} \tag{6.15}$$

where is the feature vectors derived from (Q_i, A_i) for each i = 1,2,...,n and is the feature vectors derived from U_t. Based on the confidence of the similarity, the best describes the context information, the top K matched questions, and corresponding answers are obtained.

6.4 Results and Discussion

The solution is deployed in support ticketing system for large IT services to automate the user queries/questions based on leave-, attendance-, salary-, travel-, and finance-related queries. We have captured all the question answers related to leave, salary, travel, etc. into the systems and performed steps given in Sect. 6.3. The system can predict the answer to the user query/question using question answering model. We can automate more than 80% of user queries using this solution which can reduce manual assistance to a large extent. The deployed ticketing solution was built on close to 10k question-answer pairs and was tested for more than 1 lakh user utterances. We have achieved an accuracy of about 85% while addressing the user queries. The accuracy number is based on the user satisfaction level (yes/no), which was provided by each user just after receiving the response.

The sample utterance for question and answers and words which are used for feature vector computation are given in below tables as examples. Let the user utterance related to paternity leave be:

1. Is paternity leave subject to manager's approval, and users also asked the same user utterances in different way such as:
2. paternity leaves require manager approval?
3. any pointers on paternity approval-related stuff?
4. after applying paternity leave, whether it will go for any approval?
5. what is the process to apply paternity leave?

Table 6.1 Support ticketing system: question and answers capture for training

Question	Answer
Is paternity leave subject to manager's approval	Paternity leave is an auto-approved leave; however your manager gets an intimation of the same
What is the vertical level travel or FTR	Dear User, You cannot raise the FTR/ foreign travel request for vertical level. FTR is always project based
I am trying to raise the amendment but getting the error as reservation is mandatory for raising amendment	Employees have to be reserved for an onsite US indent before raising the amendment; it's a valid message

Table 6.2 Support ticketing system: Spanish question and answers capture for training

Question	Answer
Quin vende la oferta de Crdito de Consumo del programa de ahorro ms crdito?	Quin vende la oferta de Crdito de Consumo del programa de ahorro ms crdito?, Quin puede vender la oferta del Crdito de Consumo de ahorro ms crdito?
Tiene Comisin la Cartola Solicitada por Meson para la Cuenta de Ahorro Pensin Alimenticia?	Tiene Comisin la Cartola Solicitada por Meson para la Cuenta de Ahorro Pensin Alimenticia?, La cartola de Ahorro Pensin Alimenticia sin libreta tiene comisin?, Cul es el valor de la comisin de la cartola de Ahorro Pensin Alimenticia?
Una cuenta de Ahorro Pensin Alimenticia, puede cambiarse a sin libreta?	Una cuenta de Ahorro Pensin Alimenticia, puede cambiarse a sin libreta?, Cmo se puede obtener la tarjeta para ATM de una cuenta de Ahorro Pensin Alimenticia?, Una Cuenta de Ahorro Pensin Alimenticia con libreta puede cambiarse a sin libreta?

These variations are processed through steps given in Sect. 6.3 and arrive at the unique matching question given in Table 6.1, and corresponding answer is retrieved to the user.

Unlike English, Spanish and French languages are morphologically rich in nature where the inflections play a vital role in building the sentences/questions. Moreover, while preprocessing, the stopwords and special characters removal are difficult in the languages other than English. Thus building a generic module for certain NLP tasks is difficult. Thus, we used language-specific components for each language separately. Table 6.2 shows the Spanish dataset used for training.

6.5 Conclusion

The solution proposed in this paper discusses the semantic model for question answering utterances. This model built based on pre-trained spaCy and GloVe vectors. The cosine similarity distance measure helps in matching the user query with the question answering utterances. The proposed solution is deployed in port ticketing system for large IT industry to automate user queries/utterances by in large. The solution is able to capture more than 85% of accuracy on testing datasets of user queries. In the future, we want to extend this solution to resolve user queries on any other domains such as health care, finance, automobile, etc.

References

1. Li H, Xu J (2013) Semantic matching in search. Found Trends Inf Retr 7:343–469
2. Berger A, Caruana R, Cohn D, Freitag D, Mittal V (2000) Bridging the lexical chasm: statistical approaches to answer-finding. In: Proceedings of the 23rd Annual International ACM SIGIR Conference on Research and Development in Information Retrieval, Athens, Greece, pp 192–199
3. Bengio Y, Ducharme R, Vincent P, Janvin C (2003) A neural probabilistic language model. J Mach Learn Res 3:1137–1155
4. Mikolov T, Sutskever I, Chen K, Corrado GS, Dean J (2013) Distributed representations of words and phrases and their compositionality. In: Advances in Neural Information Processing Systems (NIPS), pp 3111–3119
5. Suyog T, Gopichand A, Balaji J, Pandurang N (2018) A question answering model based on semantic matcher for support ticketing system. In: International Conference on Advances in Computing and Data Sciences (ICACDS), 2018 Publishing in Springer Communications in Computer and Information Science (CCIS) Series (Accepted and Yet to Publish)
6. Shen Y, Rong W, Jiang N, Peng B, Tang J, Xiong Z (2017) Word embedding based correlation model for question/answer matching. In: Proceedings of the Thirty-First AAAI Conference on Artificial Intelligence (AAAI-17), CA, USA, pp 3511–3517
7. Kutuzov A, Kuzmenko E (2016) Neural embedding language models in semantic clustering of web search results. In: Proceedings of the Tenth International Conference on Language Resources and Evaluation (LREC'16), Portorož, Slovenia, pp 3044–3048
8. Wan S, Lan Y, Guo J, Xu J, Pang L, Cheng X (2016) A deep architecture for semantic matching with multiple positional sentence representations. In: Proceeding AAAI'16: Proceedings of the Thirtieth AAAI Conference on Artificial Intelligence, AZ, USA, pp 2835–2841
9. Grbovic M, Djuric N, Feng A, Ordentlich E (2016) Scalable semantic matching of queries to Ads in sponsored search advertising. In: Proceeding SIGIR'16: Proceedings of the 39th International ACM SIGIR Conference on Research and Development in Information Retrieval, Tuscany, Italy, pp 375–384
10. Guo J, Fan Y, Ai Q, Croft WB (2016) Semantic matching by non-linear word transportation for information retrieval. In: Proceeding CIKM'16: Proceedings of the 25th ACM International on Conference on Information and Knowledge Management, Indianapolis, IN, USA, pp 701–710
11. Molino P, Aiello LM (2014) Distributed representations for semantic matching in non-factoid question answering. In: Proceedings of Workshop on Semantic Matching in Information Retrieval Co-located with the 37th International ACM SIGIR Conference on Research and Development in Information Retrieval, SMIR@SIGIR'14, Gold Coast, QLD, Australia, pp 38–45
12. Giordani A, Moschitti A (2009) Semantic mapping between natural language questions and SQL queries via syntactic pairing. In: Proceedings of the 14th International Conference on Applications of Natural Language to Information Systems, NLDB'09, Saarbrucken, Germany, pp 207–221

13. Levy O, Goldberg Y (2014) Dependency-based word embeddings. In: Proceedings of the 52nd Annual Meeting of the Association for Computational Linguistics (vol 2: Short Papers), Baltimore, pp 302–308
14. Pennington J, Socher R, Manning CD (2014) GloVe: global vectors for word representation. In: Proceedings of the 2014 Conference on Empirical Methods in Natural Language Processing, Doha, Qatar, pp 1532–1543

Part III
Natural Computing for Unsupervised Learning

Chapter 7
Chemical Reaction-Based Optimization Algorithm for Solving Clustering Problems

Yugal Kumar, Neeraj Dahiya, Sanjay Malik, Geeta Yadav, and Vijendra Singh

7.1 Introduction

Clustering is an important data analysis and well-recognized technique in the field of data mining. Since past few decades, it has attained a wide awareness from research community having theoretical as well as practical views. It is an unsupervised classification technique, and there is no need for training and testing of data objects. This technique can split the data objects into different clusters using a distance criterion. Literature survey has revealed the use of this technique in diverse research domains [3, 4, 7, 8, 18, 19, 21] . In recent years, a number of heuristics algorithms have been reported to determine the optimal solution for clustering problems. These algorithms have been derived from the heuristics, natural phenomena, swarms behavior, insects, animals, etc. Some physics-based algorithms (such as charged system search algorithm, galaxy-based algorithm, black hole optimization algorithm, magnetic charged system search algorithm) have also been used to find out the optimal solution for optimization problems [6, 9, 10, 16]. Some other algorithms like CSS [11], MCSS [12], BH [6], and CSO [13, 14] have also been applied successfully to solve the clustering problems. With time, the

Y. Kumar (✉)
Department of Computer Science and Engineering, JUIT, Waknaghat, Himachal Pradesh, India

N. Dahiya · S. Malik
Department of Computer Science and Engineering, SRM University, Delhi-NCR Campus, Ghaziabad, Uttar Pradesh, India

G. Yadav
Department of Pharmacy, Manav Bharti University, Solan, Himachal Pradesh, India

V. Singh
Department of Computer Science and Engineering, The Northcap University, Gurugram, India

X. Li, K.-C. Wong (eds.), *Natural Computing for Unsupervised Learning*, Unsupervised and Semi-Supervised Learning,
https://doi.org/10.1007/978-3-319-98566-4_7

importance and significance of these algorithms have increased among the research community due to following reasons:

- These algorithms have a variety of solution strategies.
- These algorithms do not depend on the size of the problem, variables, constraints, and solution space.
- These algorithms can be adopted according to the problem. Only need to adjust the objectives and fitness functions. It means no hard mathematical functions are defined in these algorithms.
- These algorithms provide efficient and effective results for combinatorial and nonlinear problems.
- These algorithms have the low computation power and minimum chances to trap in local optima.
- These algorithms have the capacity to solve the problem with different initial points.
- These algorithms provide better results as compared to the classical methods.

Due to abovementioned advantages, these algorithms have been used in diverse fields such as image processing, biology, sciences, market research, biomedical process, engineering, etc. Some new versions of these algorithms also have been proposed. Artificial chemical reaction optimization (ACRO) algorithm is a chemistry-based heuristic algorithm to solve the global optimization problems [1, 2]. This algorithm is based on the chemical reactions such as their types, evaluation, decomposition, etc. ACRO algorithm is mainly described in terms of reactants. For reactants, any encoding schemes, like integer, real, floating, and character, are used for solving the optimization problems. Further, in the ACRO algorithm, some reaction rules are specified to consume and generate the reactants. The execution of the algorithm is stopped when termination condition is met. The primary objective of this work is to investigate the competences of the ACRO algorithm for finding the optimal cluster centers for partitional clustering problems. The performance of proposed algorithm is examined on well-known benchmark datasets downloaded from UCI repository and few artificial datasets. The simulation results are compared with several meta-heuristic algorithms such as K-means [15], particle swarm optimization (PSO) [20], ant colony optimization (ACO) [17], and bat algorithm (BA) [5].

Organization of the Paper: The ACRO algorithm and different chemical reactions adopted in the algorithm are presented in Sect. 7.2. Section 7.3 presents the ACRO algorithm for partitional clustering problems. Section 7.4 illustrates the results of the proposed study. Section 7.5 gives a summary of the entire work.

7.2 Chemical Reactions

The chemical reaction can be defined as the formation of new chemical products using chemical reactants. Reaction completion timings may differ like some chemical reactions take few minutes, while others may take hours or days. Some chemical reactions require single step for product formation, while others need multiple steps. In general, chemical reactions can be divided into two categories such as consecutive reaction and competing reaction. Basically, the motion of the electrons is generally responsible for the formation and breakage of the chemical bond in chemical reactions. It is also possible that the newly derive products, may differ in properties from reactants. A chemical process contains different types of molecules or chemical species, and further, these molecules can participate in one or more chemical reactions. Reactions may be exothermic or endothermic in nature. Reactions may be reversible or irreversible. In some reactions, the rate of the forward reaction is equal to rate of the reverse reaction such condition is known as chemical equilibrium condition. In equilibrium state, concentrations of reactants and products are constant. Observation shows that sometime output of a reaction is the input of other reaction. Broadly, chemical reactions can be characterized into following categories:

7.2.1 Synthesis Reactions

In these reactions, two or more reactants combine for producing a single product or compound. The reaction is given below.

$$2\mathrm{K} + \mathrm{Cl}_2 \rightarrow 2\mathrm{KCl} \qquad \text{(Elements)}$$

$$2\mathrm{KCl} + 3\mathrm{O}_2 \rightarrow 2\mathrm{KClO}_3 \qquad \text{(Compounds)}$$

7.2.2 Decomposition Reactions

Through this reaction, two or more compounds form using a single reactant. It just reverses of the synthesis reaction. Decomposition reactions need some energy source like heat, light, or electricity. A decomposition reaction can be shown as

$$\mathrm{MgCl}_2 \rightarrow \mathrm{Mg} + \mathrm{Cl}_2$$

7.2.3 Single Displacement Reactions

In single displacement reaction, an element can interact with another element of a compound and takes position of another element in the resultant compound. This reaction can be given as

$$\mathrm{Zn} + 2\mathrm{HCl} \rightarrow 2\mathrm{ZnCl}_2 + \mathrm{H}_2$$

7.2.4 Metathesis

In metathesis reactions, after the interaction of the elements, the position of cations and anions changes and produce two new elements. This reaction can be specified as given below.

$$AgNO_3 + NaCl \rightarrow 2AgCl + NaNO_3$$

7.2.5 Combustion Reactions

In this reaction, oxygen reacts with an element and produced energy in terms of heat and light. Combustion reactions can be given as

$$C_2H_5 + 3O_2 \rightarrow 2CO_2 + 3H_2O$$

7.2.6 Redox Reactions

In this reaction, electrons transfer from one reactant to another reactant in terms of release and uptake. The element that can uptake the electrons is called an oxidizing agent, whereas the element that can release the electrons is known as the reducing agent. The example of such reaction is given as

$$MnO_4^- \,(aq) + I^- \,(aq) \rightarrow Mn^{+2} \,(aq) + I_2 \,(s)$$

Mn gains five electrons to form Mn^{2+}. Further, iodine lost electrons, so it is an oxidizing agent. Moreover, Mn^{2+} gains five electrons, so it is a reducing agent.

7.2.7 Reversible Reactions

These reactions occur both in forward and backward direction. In such types of reactions, the products change into reactants and reactants into products. The example of such reaction is given below.

$$H_2CO_3 \,(l) + H_2O \,(l) \leftrightarrow HCO_3{}^- \,(aq) + H_3O \,(aq)$$

7.3 Artificial Chemical Reaction Optimization (ACRO) Algorithm

The motivation behind the artificial chemical reaction optimization (ACRO) algorithm is the chemical reaction. It is discovered that atoms and molecules constantly move and collide in a viscous fluid filled 2D cellular space. Atoms are defined as the elementary particle of a chemical reaction and having a type, charge, mass, position, radius, orientation, and velocity. Whereas, a molecule can be defined as a group of two or more atoms linked through bonds. During a chemical reaction, several changes occur like formation and breaking of bonds, orientation change, substitution, and replacement in molecules. It is considered that ACRO algorithm begins with the action of reactants in a vessel. Let us consider a fixed volume vessel filled with a uniform mixture of N chemical reactants which undergo different types of chemical reactions. Let $Ri \in (1 \leq i \geq N)$ is the list of chemical types, and it is considered that these types can cooperate through M specified chemical reaction channels. In ACRO algorithm, the encoding scheme of the reactants depends on the user choice, and it can be binary, real, string, integer, etc. The new reactant produces through the interaction of one or two reactants. The ACRO algorithm starts with a set of initial reactants in a solution. Further, reactants are consumed and produced through different chemical reactions. The algorithm stops; either the termination condition is met or when no more reaction occurs. For the chemical reactions, reactants are chosen based on their concentrations and potentials. Moreover, it is observed that consecutive and competing reactions are the two mainly reported reactions. Reactants are joined together serially in a consecutive reaction, whereas in competing type reactions, different products are produced depending on the specific condition. It is also noticed that sometimes output of one reaction may act as a reactant for other reaction. There are many factors that can affect the execution of a reaction. However, the ACRO algorithm is based on a simple concept of an equal probability of monomolecular or bimolecular reactions and their alternatives. The main steps of the ACRO algorithm are listed below.

Step 1: Initialize the problem and algorithm parameters.
Step 2: Evaluate the initial reactants and evaluation.
Step 3: Apply chemical reactions.
Step 4: Update reactants.
Step 5: Termination condition.

7.4 Proposed ACRO Algorithm for Clustering Problems

This section describes the proposed ACRO algorithm to find the optimal cluster centers for the partition-based clustering problems. In partition-based clustering problems, numbers of clusters are known in advance, and the main task is to find the optimum set of cluster centers such that the intra-cluster distance between cluster centers and objects should be minimized. The steps of the proposed algorithm are described in Sect. 7.4.1.

7.4.1 Steps of Proposed ACRO Algorithm

Step 1: Initialization of the User-Defined Parameters and Problem Statement
The clustering problem can be expressed as minimization problem and defined as follows:

$$\text{Minimize } f\left(x \textbf{ subject } to\ x_i \in C_k; \textbf{ where } i = 1, 2, 3, \ldots, D \textbf{ and } k = 1, 2, \ldots, K\right)$$

where $f(x)$ is an objective function which is defined in terms of Euclidean distance, $x_i$is the set of data objects, D is the dimension of data object, C_k is the set of kth cluster centers, and K describes a number of clusters in a given dataset. It is to be noted that the computed cluster centers should be in the range of $x_i^{\min}$and $x_i^{\max}$and also known as boundary constraints. Further, in this work, integer encoding scheme is used to obtain the desired results. The user-defined parameters of ACROA and Reactnum (maximum number of iterations) are also defined in this step.

Step 2: Set Initial Reactants and Evaluation
In this step, the population of the ACRO algorithm is specified; the population is defined in terms of initial reactants. The initial reactants are uniformly identified from the feasible search space. For clustering problem, initially, two reactants, i.e., R_1 and R_2, are identified from the dataset such as $R_1 = \{x_{i,\,1}, x_{i,\,2}, \ldots, x_{i,\,d}\}$, $R_2 = \{x_{j,\,1}, x_{j,\,2}, \ldots, x_{j,\,d}\}$, where d denotes the length (dimension) of reactant. The number of reactants is similar to a number of clusters (K). Then, rests of reactants are derived from initially determined reactants R_1 and R_2 using the following procedure. To compute R_1 and R_2, firstly, a dividing factor (k) is initialized. Suppose $k = 2$; then, two new extra reactants, i.e., R_3, R_4, are generated from R_1 and R_2 using below mentioned procedure:

$$R_3 = \left\{ \begin{array}{c} r * x_{i,1}, r * x_{i,2}, \ldots, r * x_{i,\frac{d}{2}}; \\ r * x_{j,\frac{d}{2}+1}, r * x_{j,d-\frac{1}{2}+2}, \ldots, r * x_{j,1} \end{array} \right\} \tag{7.1}$$

$$R_4 = \left\{ \begin{array}{c} r * x_{j,1}, r * x_{j,2}, \ldots, r * x_{j,\frac{d}{2}}; \\ r * x_{i,\frac{d}{2}+1}, r * x_{i,\ d-\frac{1}{2}+2}, \ldots, r * x_{i,1} \end{array} \right\} \tag{7.2}$$

where r denotes a random number and should be in the range of $0 \leq r \geq 1$. Further, more reactants are generated using the following procedure until the number of reactants is not similar to the desired clusters (K):

$$R_5 = \left\{ \begin{array}{c} r * x_{i,1}, r * x_{i,2}, \ldots, r * x_{i,\frac{2d}{3}}; \\ r * x_{j,2d/3+1}, r * x_{j,2(d-1)/3+2}, \ldots, r * x_{j,1} \end{array} \right\} \tag{7.3}$$

$$R_6 = \left\{ \begin{array}{c} r * x_{i,1}, r * x_{i,2}, \ldots, r * x_{i,\frac{d}{3}};\ r * x_{j,d/3+1}, r * x_{j,2(d-1)/3+2}, \\ r * x_{i,\frac{2d}{3}}, \ldots, r * x_{j,1} \end{array} \right\} \tag{7.4}$$

$$R_7 = \left\{ r * x_{i,1}, r * x_{i,2}, \ldots, r * x_{i,\frac{d}{3}};\ r * x_{j,d/3+1}, , \ldots, r * x_{j,1} \right\} \tag{7.5}$$

$$R_8 = \left\{ r * x_{j,1}, r * x_{j,2}, \ldots, r * x_{j,\frac{d}{3}};\ r * x_{i,\frac{d}{3}},\ \ldots, r * x_{i,1} \right\} \tag{7.6}$$

$$R_9 = \left\{ \begin{array}{c} r * x_{j,1}, r * x_{j,2}, \ldots, r * x_{j,\frac{d}{3}};\ r * x_{i,\frac{d}{3}+1}, r * x_{i,\frac{2(d-1)}{3}}, \\ r * x_{j,2d/3+1} \ldots, r * x_{i,1} \end{array} \right\} \tag{7.7}$$

$$R_{10} = \left\{ r * x_{j,1}, r * x_{j,2}, \ldots, r * x_{j,\frac{2d}{3}}; r * x_{i,\frac{2(d-1)}{3}+1}, r * x_{i,\frac{2d}{3}+2}, \ldots, r * x_{i,1} \right\} \tag{7.8}$$

Step 3: Applying Chemical Reactions
Step 3.1: Bimolecular Reactions

Let us consider $R_1 = \{x_{i,1}, x_{i,2}, \ldots, x_{i,d}\}$ and $R_2 = \{x_{j,1}, x_{j,2}, \ldots, x_{j,d}\}$ are two reactants that can participate in bimolecular reaction. As a result of this, various bimolecular reactions are incorporated in the ACRO algorithm such as synthesis reaction, displacement reaction, redox 2 reaction, and monomolecular reaction. For all these operations, integer representation of the population is considered rather than binary encoding. The detailed explanation of these reactions is described as below.

Step 3.2: Synthesis Reaction: Using this reaction, a new reactant can be obtained using the following equation:

$$R = \left(r_1, r_2, r_3, \ldots, r_i, r_j, r_k, \ldots, r_n\right); \text{where, } R_{i,\text{new}} = R_i + \lambda_i \left(R_j - R_i\right) \tag{7.9}$$

In the above equation, λ_i is a random value in the interval [0.25, 1.25]; R_i and R_j are randomly selected reactants.

Step 3.3: Displacement Reaction

The new reactants are obtained using below mentioned procedure. Suppose $R_k = (R_1, R_2 \ldots R_i, R_j, \ldots, R_K)$ where, $k = 1,2 \ldots \ldots K$:

$$\mathrm{R}_{\mathrm{i,new}} = R_i \left(1 - \lambda_{\mathrm{td}} R_j\right) \tag{7.10}$$

$$R_{j,\text{new}} = \lambda_{\mathrm{td}} R_j + (1 - \lambda_{\mathrm{td}} R_i) \tag{7.11}$$

where $\lambda_{td} \in \{0, 1\}$ and$\lambda_{td+1} = 2.3\ (\lambda_{td})^{2\sin(\pi\lambda_{td})}$ and where in λ_{td} the suffix "td" is increased by 1, when the reaction is completed.

Step 3.4: Redox 2 Reaction

In this reaction, if R_1 is a better reactant in terms of the objective function, then in $\lambda_{tr,}$ the suffix "tr" is updated by using 1 when the reaction is performed and it is computed using the following equation:

$$R_{\mathrm{i,new}} = R_i + \lambda_{\mathrm{tr}}\left(R_i - R_j\right);\ \text{where,}\quad \lambda_{\mathrm{tr}} \in \{0, 1\} \tag{7.12}$$

Otherwise, if R_2 is a better reactant, then in $\lambda_{tr,}$ the suffix "tr" is updated by using 1 when the reaction is performed and it is computed using the following equation:

$$R_{j,\mathrm{new}} = R_j + \lambda_{\mathrm{tr}}\left(R_j - R_i\right);\ \text{where,}\quad \lambda_{\mathrm{tr}} \in \{0, 1\} \tag{7.13}$$

The value of λ_{tr} is computed using the following equation:

$$\lambda_{\mathrm{tr}+1} = \begin{cases} 0 & \lambda_{\mathrm{tr}} = 0 \\ \frac{1}{\lambda_{\mathrm{tr}}\ \mathrm{mod}(1)}\lambda_{\mathrm{tr}} & \in (0, 1) \end{cases} \tag{7.14}$$

$$1/\lambda_{\mathrm{tr}}\ \mathrm{mod}(1) = \frac{1}{\lambda_{\mathrm{tr}}} - \left\lfloor \frac{1}{\lambda_{\mathrm{tr}}} \right\rfloor \tag{7.15}$$

Step 3.4: Monomolecular Reactions

Step 3.4.1: Decomposition Reaction

In this reaction, suppose $R = (R_1, R_2 \ldots R_i, R_j, \ldots, R_n)$ is the reactant and $C_k \in \{R_m, R_n\}$ is an atom that takes part in monomolecular reaction. The new atom of the molecule $R_{i,\text{ new}}$ is a random population or reactant from $\{R_m, R_n\} \in C_k$, but $\{R_m, R_n\} \neq (R_i$ and $R_j)$.

Step 3.4.2: Redox 1 Reaction

In this reaction, the new reactant is generated using the following procedure:

$$R_{i,\mathrm{new}} = R_m + \lambda_t\,(R_n - R_m) \tag{7.16}$$

where $\lambda_t \in \{0, 1\}$ such that initial $\lambda_0 \in \{0.0, 0.25, 0.5, 0.75, 1.0\}$ and $\lambda_{t+1} = 4\lambda_t(1 - \lambda_t)$; t is updated by using 1 when the reaction is performed.

Step 4: Reactants Update

In this step, a chemical equilibrium test is carried out. If the fitness function of the new generated reactants is better than another, the generated reactant includes into the chemical reaction process and excludes the worst reactant from reversible chemical reactions.

Table 7.1 Description of the datasets used for experiment

Dataset	Cluster (K)	Attributes	Objects	Objects in clusters
Iris	3	4	150	(50, 50, 50)
Wine	3	13	178	(59, 71, 48)
CMC	3	9	1473	(629,334, 510)
Cancer	2	9	683	(444, 239)
Glass	6	9	214	(70,17, 76, 13, 9, 29)

Step 5: Termination Condition
If the Reactnum is equal to the user-defined maximum number of iterations, then the algorithm stops its execution and produces the optimal cluster centers; otherwise steps 3 and 4 are repeated, until the desired results are not obtained or termination condition is not satisfied.

7.5 Simulation Results

This section presents the experimental results of our study. To examine the efficacy of the proposed ACRO based clustering algorithm, some well-known benchmark datasets are taken from the UCI repository. These datasets are iris, CMC, glass, wine, and cancer. The description these datasets are reported in Table 7.1. Further, in this work, intra-cluster distance and f-measure parameters are adopted as performance measure parameters. The intra-cluster distance parameter presents the quality of clusters, and it can be measured using best, average, and worst intra-cluster distances. The precision can be computed using f-measure parameter. The MATLAB environment is used to implement the proposed algorithm using Windows-based operating system. Table 7.1 demonstrates the details of datasets taken for the experiments. The simulation results of proposed algorithm are compared with other well-known clustering algorithms K-means [15], particle swarm optimization (PSO) [20], ant colony optimization (ACO) [17], and bat algorithm (BA) [5].

7.5.1 Performance Matrices

In this section, different performance matrices are described that are adopted to compute the performance of the ACRO algorithm for partitional clustering problems. In this work, f-measure and intra-cluster distance parameters are taken to evaluate the performance of the algorithms. These matrices are explained as below.

- Intra-cluster distances

This parameter is used to compute the distance between data objects and cluster centers and also represents the quality of clustering. The results describe using best, average, and worst cases.

- F-Measure

It is expressed in terms of recall and precision. The value of f-measure parameter is computed through Eq. 7.17.

$$F\left(i, j\right) = \sum\nolimits_{i=1}^{n} \frac{n_i}{n}^{*} \max_{i}{}^{*} F\left(i, j\right) \tag{7.17}$$

Where, $F(i,j)$is determined using Eq. 7.18.

$$F\left(i, j\right) = \frac{2 * (\text{Recall} * \text{Precision})}{(\text{Recall} + \text{Precision})} \tag{7.18}$$

The recall and precision are measured using Eq. 7.19.

$$\text{Recall} \left(r\left(i, j\right)\right) = \frac{n_{i,j}}{n_i} \quad \text{and} \quad \text{Precision} \left(p\left(i, j\right)\right) = \frac{n_{i,j}}{n_j} \tag{7.19}$$

7.5.2 *Results and Discussion*

This subsection describes the results of our study using proposed algorithm. Tables 7.2 and 7.3 present the results of the proposed ACRO algorithm and other clustering using artificial dataset 1 and artificial dataset 2. It is noticed that the proposed algorithm gives better results in comparison to all other algorithms. It is also observed that K-means algorithm obtains worst results among all other algorithm using intra-cluster distance parameter both of artificial dataset1 and artificial dataset 2 using intra-cluster distance parameter.

Tables 7.4 and 7.5 illustrate the results of the proposed algorithm and other algorithms using iris and cancer dataset. It is reported that the proposed algorithm

Table 7.2 Performance comparison of proposed ACRO and other algorithms using artificial dataset 1

Parameters	K-means	PSO	ACO	CSO	BA	Proposed ACRO
Best case	161.32	153.45	154.29	159.73	151.81	143.36
Avg. case	166.12	161.24	158.52	163.37	154.98	149.56
Worst case	174.64	169.39	165.42	168.13	159.08	154.21
SD	7.625	6.437	4.712	5.418	4.564	5.243
F-Measure	0.94	0.96	0.99	0.95	0.97	1

Table 7.3 Performance comparison of proposed ACRO and other algorithms using artificial dataset 2

Parameters	K-means	PSO	ACO	CSO	Bat	Proposed ACRO
Best case	761.45	753.24	759.87	752.34	756.09	746.53
Avg. case	768.38	759.82	766.15	756.21	761.44	752.26
Worst case	776.49	769.58	773.41	764.39	169.17	758.13
SD	6.837	7.614	6.845	4.936	5.432	4.214
F-Measure	0.89	0.91	0.93	0.937	0.946	0.976

Table 7.4 Performance comparison of proposed ACRO and other algorithms using iris dataset

Parameters	K-means	PSO	ACO	CSO	Bat	Proposed ACRO
Best case	97.52	97.05	97.21	96.98	96.84	95.56
Avg. case	113.56	98.73	98.36	97.64	97.53	96.73
Worst case	125.23	99.89	99.59	98.78	98.09	97.48
SD	15.326	0.467	0.426	0.392	0.263	0.196
F-Measure	0.781	0.78	0.778	0.781	0.782	0.785

Table 7.5 Performance comparison of proposed ACRO and other algorithms using cancer dataset

Parameters	K-means	PSO	ACO	CSO	Bat	Proposed ACRO
Best case	2989.46	2978.68	2983.49	2985.16	2972.36	2912.66
Avg. case	3248.25	3116.64	3178.09	3124.15	3098.93	3063.34
Worst case	3566.94	3358.43	3292.41	3443.56	3282.75	3179.25
SD	256.58	107.14	93.45	128.46	56.24	71.22
F-Measure	0.832	0.826	0.829	0.831	0.833	0.835

obtains better quality results than other algorithms. Further, it is noted that the K-means algorithm obtains a maximum intra-cluster distance among all algorithms using both of datasets. It is also noticed that the performance of the K-means, PSO, ACO, and CSO algorithm is similar in case of f-measure parameter using iris dataset. But, the significant difference occurs in terms of intra-cluster distance parameter. It is observed that the PSO algorithm has worst performance using f-measure parameter among the rest of algorithms for cancer dataset.

Tables 7.6, 7.7, and 7.8 demonstrate the results of proposed ACRO and other algorithms using CMC, wine, and glass dataset. It is revealed that the proposed algorithm provides enhanced results in comparison to other algorithms for CMC and wine datasets. Moreover, it is noticed that the performance of BA algorithm is better for the glass dataset in comparison to all other algorithms. Further, it is reported that K-means algorithm obtains maximum intra-cluster distance for CMC and wine datasets, whereas, an ACO algorithm obtains a maximum intra-cluster distance for glass dataset. For wine and glass datasets, K-means and ACO algorithms exhibit worst f-measure results. In case of the CMC dataset, PSO, ACO, and CSO provide similar f-measure results.

Table 7.6 Performance comparison of proposed ACRO and other algorithms using CMC dataset

Parameters	K-means	PSO	ACO	CSO	Bat	Proposed ACRO
Best case	5828.25	5792.48	5756.42	5712.78	5689.16	5681.56
Avg. case	5903.82	5846.63	5831.25	5804.52	5778.14	5746.32
Worst case	5974.46	5936.14	5929.36	5921.28	5914.25	5894.63
SD	49.62	48.86	44.34	43.29	39.54	36.41
F-Measure	0.337	0.333	0.332	0.334	0.336	0.339

Table 7.7 Performance comparison of proposed ACRO and other algorithms using wine dataset

Parameters	K-means	PSO	ACO	CSO	Bat	Proposed ACRO
Best case	16768.18	16483.61	16448.35	16431.76	16372.02	16256.42
Avg. case	18061.24	16417.47	16530.53	16395.18	16357.89	16336.21
Worst case	18764.49	16594.26	16616.36	16589.54	16556.76	16396. 56
SD	796.13	88.27	48.86	62.41	41.78	37.83
F-Measure	0.519	0.516	0.522	0.521	0.523	0.526

Table 7.8 Performance comparison of proposed ACRO and other algorithms using glass dataset

Parameters	K-means	PSO	ACO	CSO	Bat	Proposed ACRO
Best case	222.43	264.56	273.22	256.53	256.47	261.47
Avg. case	246.51	278.71	281.46	264.44	261.61	266.23
Worst case	258.38	283.52	286.08	282.27	278.24	274.14
SD	18.32	8.59	6.58	15.43	7.09	8.11
F-Measure	0.426	0.412	0.402	0.416	0.430	0.428

Figures 7.1, 7.2 and 7.3 illustrate the dispersion of the data objects in artificial dataset 1 and dataset 2, respectively. These artificial datasets are generated in the MATLAB. Artificial dataset 1 is a two-dimensional dataset, whereas artificial dataset 2 is three-dimensional datasets. Figures 7.2, 7.3 and 7.4 show the clustering of the data objects into different clusters using the proposed algorithm.

Figures 7.5 and 7.6 show the clustering of the iris dataset using proposed artificial chemical reaction optimization algorithm. The data objects of iris dataset are divided into three clusters such as setosa, versicolor, and virginica. It is seen that data objects in setosa cluster are linearly separable from versicolor and virginica, while the data objects of versicolor and virginica clusters are linearly inseparable that can also affect the performance of the algorithm. But, it is observed that the proposed algorithm gives better results for all three clusters. Figure 7.5 shows the illustration of the data objects of the iris dataset using sepal width and petal width attributes, whereas Fig. 7.6 shows the illustration of data objects using petal length, sepal width, and petal width.

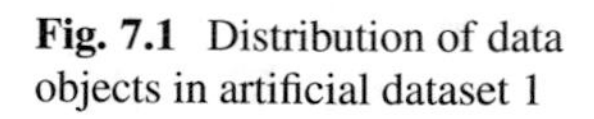

Fig. 7.1 Distribution of data objects in artificial dataset 1

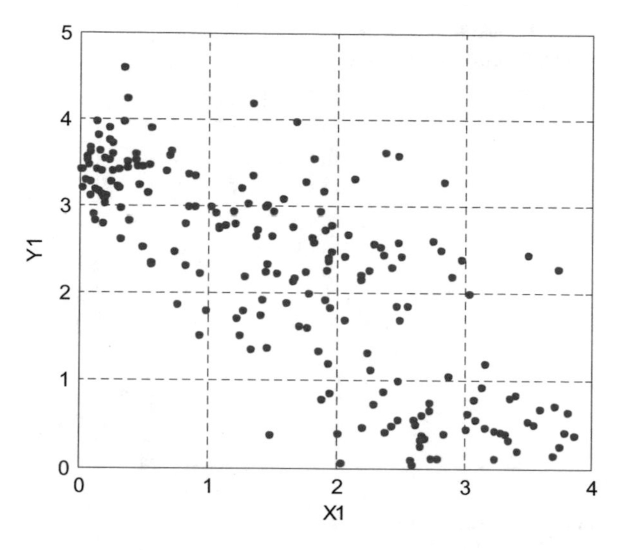

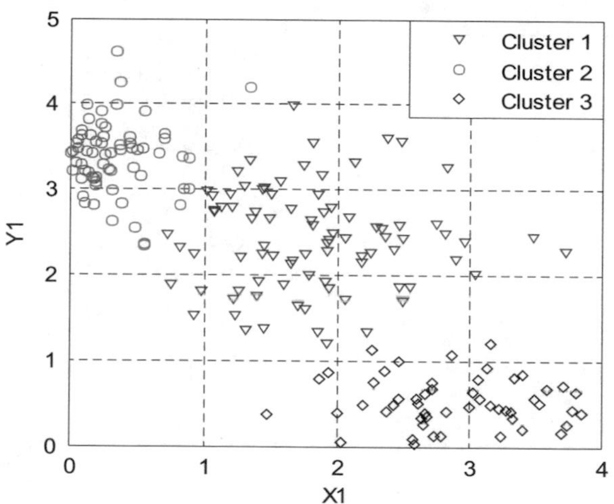

Fig. 7.2 Clustering results of proposed ACRO algorithm on artificial dataset 1

7.6 Conclusion

In this work, an artificial chemical reaction optimization algorithm is presented for solving partitional clustering problems. The proposed algorithm is inspired by the chemical reaction process. In this algorithm, reactants are used to search the optimal solution, and these reactants are uniformly determined from the search space. Further, optimal solution for the problems can be represented using the reactants. The main work of the ACRO algorithm is to measure optimal cluster centroid for partitional clustering problems. The reactants represent the initial cluster centers, which are determined uniformly from the dataset. The proposed algorithm is applied to optimize the value of initially chosen reactants through its

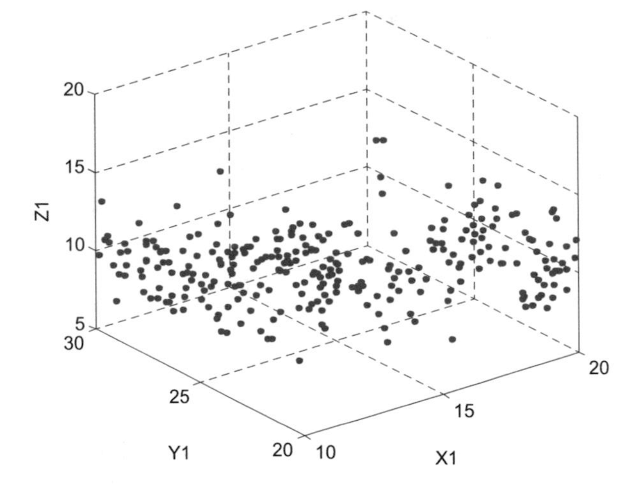

Fig. 7.3 Distribution of data objects in artificial dataset 2

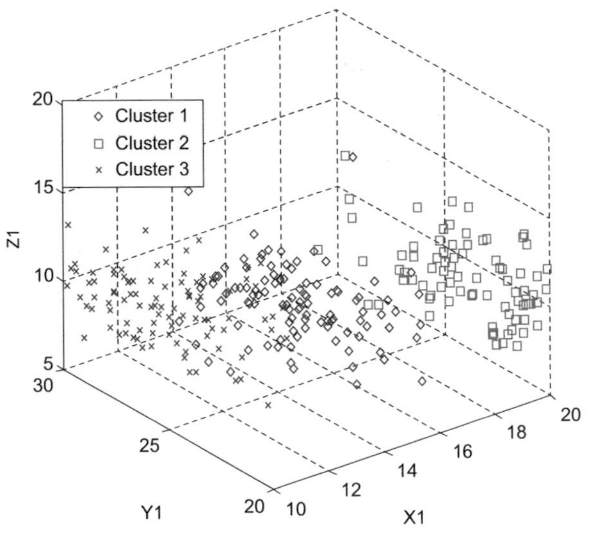

Fig. 7.4 Clustering results of proposed ACRO algorithm on artificial dataset 2

various steps. The performance of the proposed algorithm is tested on several real-life clustering problems and compared with state-of-the-art clustering algorithms. From the simulation results, it is noticed that the proposed algorithm achieves better clustering results in comparison to other clustering algorithms. Finally, it is stated that proposed ACRO algorithm is one of the efficient and effective algorithm for solving partitional clustering problems.

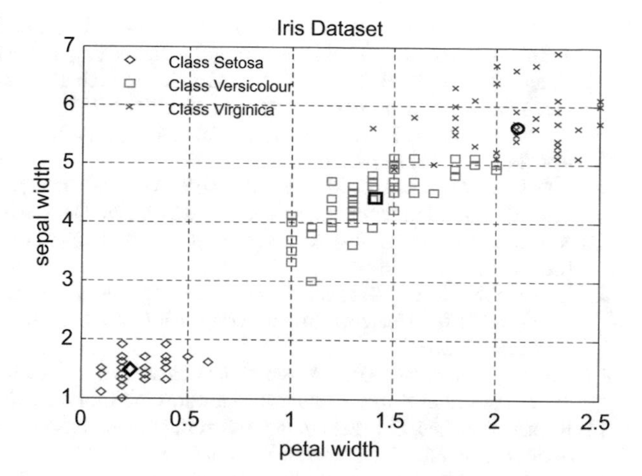

Fig. 7.5 Clustering results of proposed ACRO algorithm on iris dataset (2D view)

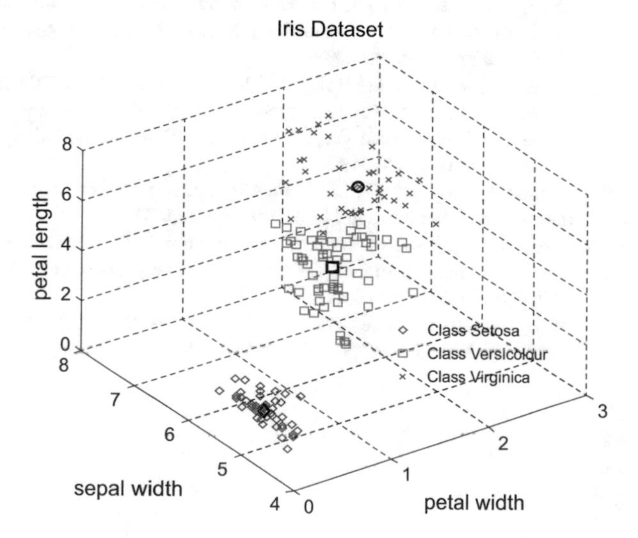

Fig. 7.6 Clustering results of proposed ACRO algorithm on iris dataset (3D view)

References

1. Alatas B (2011) ACROA: artificial chemical reaction optimization algorithm for global optimization. Expert Syst Appl 38(10):13170–13180
2. Alatas B (2012) A novel chemistry based metaheuristic optimization method for mining of classification rules. Expert Syst Appl 39(12):11080–11088
3. Anaya AR, Boticario JG (2011) Application of machine learning techniques to analyses student interactions and improve the collaboration process. Expert Syst Appl 38(2):1171–1181
4. Dunn WJ III, Greenberg MJ, Callejas SS (1976) Use of cluster analysis in the development of structure-activity relations for antitumor triazenes. J Med Chem 19(11):1299–1301
5. Gandomi AH, Yang XS, Alavi AH, Talatahari S (2013) Bat algorithm for constrained optimization tasks. Neural Comput Applic 22(6):1239–1255
6. Hatamlou A (2013) Black hole: a new heuristic optimization approach for data clustering. Inform Sci 222:175–184

7. He Y, Pan W, Lin J (2006) Cluster analysis using multivariate normal mixture models to detect differential gene expression with microarray data. Comput Stat Data Anal 51(2):641–658
8. Hung YS, Chen KLB, Yang CT, Deng GF (2013) Web usage mining for analyzing elder self-care behavior patterns. Expert Syst Appl 40(2):775–783
9. Kaveh A, Talatahari S (2010) A novel heuristic optimization method: charged system search. Acta Mech 213(3):267–289
10. Kaveh A, Share MAM, Moslehi M (2013) Magnetic charged system search: a new meta-heuristic algorithm for optimization. Acta Mech 224(1):85–107
11. Kumar Y, Sahoo G (2014) A charged system search approach for data clustering. Prog Artif Intell 2(2-3):153–166
12. Kumar Y, Sahoo G (2015) Hybridization of magnetic charge system search and particle swarm optimization for efficient data clustering using neighborhood search strategy. Soft Comput 19(12):3621–3645
13. Kumar Y, Sahoo G (2015) A hybrid data clustering approach based on improved cat swarm optimization and K-harmonic mean algorithm. AI Commun 28(4):751–764
14. Kumar Y, Sahoo G (2016) A hybridize approach for data clustering based on cat swarm optimization. Int. J Inf Commun Technol 9(1):117–141
15. MacQueen J (1967) Some methods for classification and analysis of multivariate observations. In: Fifth Berkeley Symposium on Mathematics. Statistics and Probability. University of California Press, pp 281–297
16. Shah-Hosseini H (2011) Principal components analysis by the galaxy-based search algorithm: a novel metaheuristic for continuous optimisation. Int J Comput Sci Eng 6(1-2):132–140
17. Shelokar PS, Jayaraman VK, Kulkarni BD (2004) An ant colony approach for clustering. Analy Chim Acta 509(2):187–195
18. Teppola P, Mujunen SP, Minkkinen P (1999) Adaptive Fuzzy C-Means clustering in process monitoring. Chemom Intel Lab Syst 45(1):23–38
19. Webb A (2002) Statistical pattern recognition. Wiley, New Jersey, pp 361–406
20. Zhan ZH, Zhang J, LiY CSH (2009) Adaptive particle swarm optimization. IEEE Trans Syst Man Cybern B Cybern 39:1362–1381
21. Zhou H, Liu Y (2008) Accurate integration of multi-view range images using k-means clustering. Pattern Recognit 41(1):152–175

Chapter 8
A Novel Artificial Bee Colony Algorithm for Robust Permutation Flowshop Scheduling

Shijing Ma, Yunhe Wang, and Mingjie Li

8.1 Introduction

In the past, most researchers put mass attention on the permutation flowshop scheduling problems in which the deterministic environment is considered mostly. However, in real situations, uncertain factors may arise in the permutation flowshop [1, 2]. If uncertain issues arise, the processing time of jobs may become uncertain, which will result in the variations of the finish time or the makespan of schedule. These variations may lead to an uncompleted job schedule during the expected finish time. Thus, it may affect the scheduling of other jobs and ultimately lead to the delay of products. Therefore, more attention should be paid on the study of uncertainties in flowshop scheduling.

In stochastic production environment, the job processing time is random and not known until it is accomplished. Being able to deal with these stochastic events of scheduling is pretty significant for real-world schedule systems. When the uncertain events happen, a good scheduling system should not only be able to cope with these factors but also can produce robust schedules for these situations. If the property of schedule does not dramatically degrade when the schedule faces with disruption, the schedule is a robust schedule. It means that the property of a robust schedule is not sensitive to uncertain disruptions [3]. Therefore, robust scheduling is widely utilized in various real applications, such as process industries [4], project scheduling [5],

S. Ma · Y. Wang · M. Li (✉)
College of Computer Science and Information Technology, Northeast Normal University, Changchun, China
e-mail: masj195@nenu.edu.cn; wangyh082@nenu.edu.cn

X. Li, K.-C. Wong (eds.), *Natural Computing for Unsupervised Learning*, Unsupervised and Semi-Supervised Learning,
https://doi.org/10.1007/978-3-319-98566-4_8

airline crew scheduling [6], and manufacturing industries [7]. Currently, it is very significant to study robust scheduling for manufacturing industries.

Reactive scheduling and proactive scheduling [3, 8, 9] are often used in robust scheduling. Reactive scheduling does not directly take the uncertainty into consideration during the produce of schedules but revises the schedule when disruptions or unexpected issues arise [9]. In contrast, the proactive scheduling considers the future uncertain factors in generating the initial schedule [3]. Most researchers now concentrate on the proactive scheduling for the permutation flowshop.

In terms of the quantification of the schedule robustness, different measures were proposed. The most commonly used robustness measure is to minimize the variations of a performance indicator in a schedule [3, 10–16]. The schedules based on that robustness measure do not ensure that the property indicator not surpass a certain expected value once uncertainties arise. Therefore, Daniels and Carrillo [17] and Wu, Brown, and Beck [18] proposed a novel robustness approach to maximize the probability of making sure that property indicator of a schedule not surpass the limit value. However, these researches cannot represent the practical case, because they are limited to a single machine flowshop [17, 18]. As the number of machines in a permutation increases, the impacts of uncertain factors on scheduling will rise accordingly. Liu et al. [19] firstly proposed the robust scheduling for M-machine permutation flowshop using an improved genetic algorithm to generate robust schedules. The problem in Liu et al. [19] was studied again to find better robust schedules in this chapter.

During the past several decades, different metaheuristic algorithms have been studied widely for solving scheduling problems, such as standard NEH heuristic [20] and improved genetic algorithm [19]. Artificial bee colony (ABC) algorithm was first proposed by Karaboga in 2005 for tackling continuous optimization problems. Due to the simplicity, parallelism, fast convergence, ease of implementation, and robustness, ABC has attracted a great deal of attention for solving the discrete problem, such as the permutation flowshop problem. But ABC has the shortcoming of the premature convergence and often suffers from trapping in local optimum. In order to deal with these problems, a novel ABC combined with an improved local search, which introduces a new strategy to generate solutions and a probability model to determine the acceptation of solutions, is presented in this chapter for the M-machine permutation flowshop scheduling. The comparison experiments with an improved GA and a NEH heuristic show that the novel ABC outperforms others and can produce better robust schedules for flowshop problem in rising the probability of making makespan under the expected finish time.

The rest of this chapter is organized as follows: Section 8.2 describes the M-machine permutation flowshop problem, and the robust schedule is defined in Sect. 8.3. Section 8.4 goes toward the novel artificial bee colony algorithm. The details of the experiments and comparison results are presented in Sect. 8.5.

8.2 Problem Description

In a flowshop scheduling problem, a given set of N jobs is considered to be processed on M machines. These jobs are independent and available to be processed at time zero. Each job has M operations, and each operation should be processed on a different machine. All jobs have the same processing order on every machine. The first machine has no idle time, and each machine can deal with at most one job at a time. Each job can be tackled only on one machine at a time and no preemption is allowed. Setup times and transportation times of jobs are ignored or included in the job processing times. All machines are always available when they are unused. There are sufficient buffer spaces for jobs to wait if there are no available machines. P_{ij} denotes the processing time of job i on machine j and the corresponding completion time is C_{ij}. $C_{\max}$ represents the completion time of the last job which is also called makespan. A mathematical model for this flowshop scheduling problem is as follows:

$$C_{11} = P_{11} \tag{8.1}$$

$$C_{i1} = \sum_{r=1}^{i} P_{r1}, \quad i = 1, 2, \ldots, N \tag{8.2}$$

$$C_{1j} = \sum_{r=1}^{j} P_{1r}, \quad j = 1, 2, \ldots, M \tag{8.3}$$

$$C_{ij} = \max\left\{C_{i(j-1)}, C_{(i-1)j}\right\} + P_{ij}, \quad i = 1, 2, \ldots, N \text{ and } j = 1, 2, \ldots, M \tag{8.4}$$

$$C_{\max} = C_{NM} \tag{8.5}$$

$$\text{Objective} = \max\left\{\text{Probability}\left(C_{\max} \leq \text{expected completion time}\right)\right\} \tag{8.6}$$

In real cases, the makespan of schedule may vary due to the uncertain events. Minimizing the makespan does not meet the needs of actual situation. So, in this model, the objective that maximizing the probability of making sure the makespan not surpass the expected finish time is presented as Eq. 8.6.

In order to compute makespan easily, an efficient method to calculate the makespan of schedule is introduced as follows (S_{ij} denotes the start time of job i on machine j):

1. $S_{11} = 0$.
2. For $k = 2$ to M. $S_{1k} = S_{1(k-1)} + P_{1(k-1)}$.
3. Set $i \leftarrow 2; j \leftarrow 1$.
4. $S_{ij} = S_{(i-1)j} + P_{(i-1)j}$.
5. $j \leftarrow j + 1$
6. $S_{ij} = \max\{S_{(i-1)j} + P_{(i-1)j}, S_{i(j-1)} + P_{i(j-1)}\}$.
7. If $j < M$, go back to step 5.
8. If $i = N$, stop. $C_{\max} = S_{NM} + P_{NM}$. Otherwise, set $i \leftarrow i + 1$ and $j = 1$. Go back to step 4.

8.3 Robust Schedule

Uncertain issues cause the uncertainty of processing time of jobs and then lead to the variations in makespan. For ensuring that the makespan not surpass expected finish time when there are uncertain situations, a robust schedule can be defined as follows.

For a permutation flowshop scheduling problem of jobs on machines with uncertainty of the processing time of jobs, a schedule which can give maximum probability that makespan of jobs not surpass the expected finish time is defined as robust schedule [19]. So, a permutation flowshop schedule with the objective shown in Eq. 8.6 is a robust schedule.

For robust scheduling, the uncertainty in processing time is usually represented by probabilistic technique or fuzzy logic [21]. The job processing times in fuzzy logic are supposed by fuzzy numbers if they are uncertain. Nevertheless, these fuzzy numbers are pretty hard to appoint and might be inaccurate [22]. Furthermore, fuzzy logic is computationally complicated [21]. In probabilistic technique, the processing time uncertainty is represented by its distribution. The distribution of processing times of job could be formulated through collecting and analyzing the real figure of job processing times in the permutation flowshop or by assumptions [19]. One has to consider an abundant processing time data of production to obtain the distribution of processing time. So the uncertain processing time of jobs is an independent random variable, and the central limit theorem demonstrates that the distribution formed by mass independent random variables is normally distributed. A variety of processing time data is taken into consideration to get the processing time distribution of each job, and the uncertain processing time of each job is independent random variable. Hence, the uncertain processing time of each job will follow normal distribution on the basis of the central limit theorem.

We all know that the sum of two or more independent normally distributed numbers $A \sim N(\mu_A, {\sigma_A}^2)$ and $B \sim N(\mu_B, {\sigma_B}^2)$ also follows normal distribution i.e.,

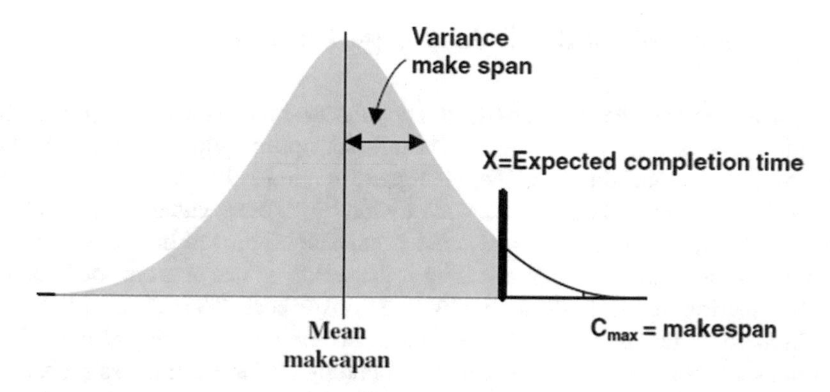

Fig. 8.1 Normal distribution curve for makespan

$A + B \sim N(\mu_A + \mu_B, \sigma_A{}^2 + \sigma_B{}^2)$ [23]. Makespan is formed by the linear combination of processing times of jobs which are normally distributed. So, the makespan is also normally distributed as demonstrated in Fig. 8.1. For normally distributed makespan $C_{\max} \sim N\left(\mu_{C_{nm}}, \sigma^2{}_{C_{nm}}\right)$, in order to compute its probability that makespan not surpass the completion time X, the calculation model is presented in Eq. 8.7.

$$\text{Probability}\,(C_{\max} \leq X) = \frac{1}{2} + \varphi(z) \tag{8.7}$$

where,

$$z \geq 0 \text{ and } z = \frac{X - \mu_{C_{NM}}}{\sigma_{C_{NM}}} \tag{8.8}$$

$$\text{and } \varphi(z) = \frac{1}{\sqrt{2\pi}} \int_{-\infty}^{z} e^{-\left(t^2/2\right)} dt \tag{8.9}$$

It is very difficult to get the exact value of the function $\varphi(z)$ indicated in Eq. 8.9. However, a novel mathematical relation to get its approximate solution is proposed by A. J. Hayter [24], which is presented in Eq. 8.10.

$$\phi(z) \approx \varphi(z) = \begin{cases} 0.1z\,(4.4 - z) & (0 \leq z \leq 2.2) \\ 0.49 & (2.2 < z < 2.6) \\ 0.50 & (z \geq 2.6) \end{cases} \tag{8.10}$$

8.4 Novel Artificial Bee Colony Algorithm (ABC)

Artificial bee colony algorithm (ABC) is a population-based stochastic optimization algorithm proposed by Karaboga and Basturk for optimization problems in 2005 [25, 26]. This method simulates the intelligent foraging behavior of the honey bee swarm. In ABC, those artificial bees are divided into three categories, which are employed bees, onlooker bees, and scout bees. The employed bees are in charge of exploiting food sources and gathering information of nectar from food sources and then sharing nectar's information with onlooker bees. The onlooker bees will wait in the hive for the nectar information to be shared by employed bees. Then each onlooker bee chooses a food source depending on the fruitfulness probability proportional of the food source to explore. Therefore, the larger the fruitfulness of food source, the more bees it attracts. Scout bees are responsible for searching new food sources in the vicinity of hive. A food source exploited fully will be abandoned by corresponding employed bee. Then the associated employed bee turns into scout bee and begins to find a new food source in the neighborhood of the hive to replace the abandoned food source. The exploitation is performed by employed and onlooker bees, while the scout bees are responsible for the exploration. The framework of the proposed ABC is shown in Fig. 8.2.

8.4.1 Encoding for Flowshop Problem

Encoding is a significant step in combinatorial optimization problems. It is a procedure that converts solution space of the problem to the search space that can be dealt with by algorithm. Good encoding can greatly improve the efficiency of algorithm. Permutation encoding is widely used in sequencing problems. In permutation encoding, every individual is represented by a string of numbers. Job-based representation is applied into the considered problem, in which the numbers and the order of numbers represent the jobs and the scheduling order of jobs, respectively. Figure 8.3 shows an individual job-based representation encoding, and ten different genes represent a schedule of ten different jobs. The number 3 assigned to the fourth gene of the individual illustrates that job 3 is the fourth job to be processed.

8.4.2 Population Initialization and Fitness Calculation

In ABC, the food source position is a potential solution for the optimization problem, and nectar amount of the food source denotes the fitness of the corresponding solution. In ABC, the initialization of the population is very important. In order to let the population have all the different possible alleles, the initial population in

```
Proposed ABC algorithm
---------------------------------------------------------------
Initialize the input parameters;
t ← 1;
For i = 1 to SN %SN denotes the population size %
 Generate a solution X_i^t randomly and evaluate f(X_i^t);
End for
---------------------------------------------------------------
Find the best solution X_best;
While stopping criteria are not true
 For i = 1 to SN % the employed bee phase%
    Select a solution X_k^t (k ≠ i) randomly and generate a new solution S_i^t
    according to X_k^t and X_i^t;
    If (f(S_i^t) > f(X_i^t))
      X_i^t ← S_i^t;
    Else
      Remember X_i^t;
    End if
  End for
  Update the X_best;
  For i = 1 to SN % the onlooker bee phase%
  Evaluate the probability value P_i of X_i^t and produce a random number
r ∈ {0,1};
  If (r < P_i)
      Select a solution X_k^t (k ≠ i) randomly and generate a new solution S_i^t
      according to X_k^t and X_i^t;
      If (f(S_i^t) > f(X_i^t))
        X_i^t ← S_i^t;
      Else
        Remember X_i^t and the search time of X_i^t plus 1;
      End if
  End if
```

Fig. 8.2 The framework of proposed ABC algorithm

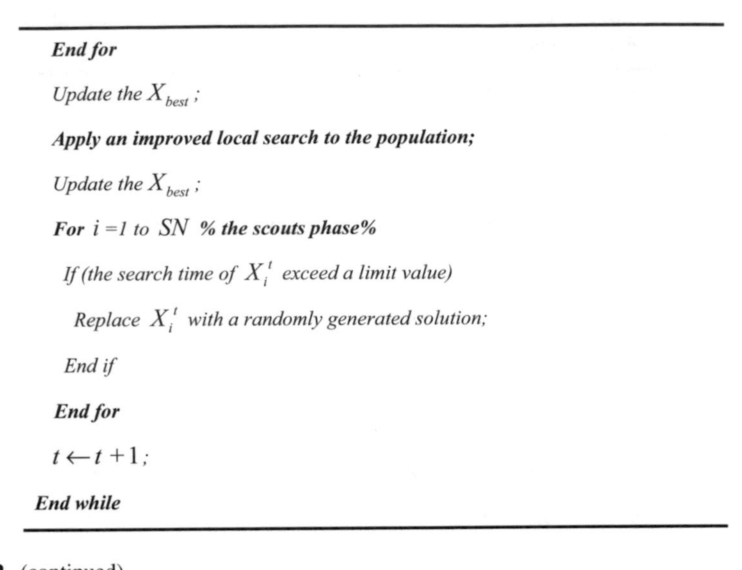

End for

Update the X_{best};

Apply an improved local search to the population;

Update the X_{best};

For $i = 1$ to SN **% the scouts phase%**

If (the search time of X_i^t exceed a limit value)

Replace X_i^t with a randomly generated solution;

End if

End for

$t \leftarrow t+1$;

End while

Fig. 8.2 (continued)

chromosome	6	2	4	3	1	9	8	10	7	5

Fig. 8.3 Job-based representation

proposed ABC is generated randomly. A better solution means the solution has the higher fitness value. In the current problem, the probability of ensuring the makespan does not surpass the expected finish time is used to determine the fitness value of each solution, which is shown in Eq. 8.11.

$$\text{fit}_i = \text{Probability}\left(C^i{}_{\max} \leq X\right) \tag{8.11}$$

where i represents the ith solution in population and X is the expected finish time.

8.4.3 Generation of a New Solution in Proposed ABC

In ABC, the bee colony contains employed bees and onlookers, and the number of employed bees (Nb) is equal to that of onlookers, which is also the same as the number of food sources (SN) [26]. At the initialization step, an initial population of the size SN will be randomly produced, in which the SN is also the size of onlookers or employed bees. And then the population (food sources) experiences repeated cycles of search procedures from those bees of three categories. First, employed bee makes a modification on a current solution based on a randomly selected solution to produce a new solution and then the fitness value of this new solution is tested.

If the new solution is better than the current one according to their fitness values, the new individual will replace the current solution. Otherwise, the bee will keep the old solution. When all employed bees finish their search process on the current population, they share the information about solutions (the food sources) with the onlookers. Next, each onlooker selects a shared solution depending on a fitness value probability proportional of the solution and exploits its vicinity to produce a new solution. If the new solution outperforms the old one, the onlooker bee will maintain the new generated solution and abandon the old one. The probability p_i is shown in Eq. 8.12.

$$p_i = \text{fit}_i / \sum_{j=1}^{\text{SN}} \text{fit}_j \tag{8.12}$$

where fit_i is the fitness value of the ith solution (associated solution) and SN is the size of the population or the food sources. In ABC, if the quality of a solution cannot be improved further after several exploration cycles of a predetermined value called "limit", the solution will be abandoned by the corresponding employed bee. And then the employed bee will become scout bee and randomly generate a new solution to replace the abandoned one.

In an N-dimensional search space, the way that employed bees and onlooker bees generate a new solution from a current one in proposed ABC is implemented through several steps as follows:

Step 1: randomly select a solution V which is different from the current one X in the population.

Step 2: randomly produce two position numbers a and b ($a < b$) which are between $[1,N]$.

Step 3: copy the substring formed by the numbers between a and b (contains a and b) of X into the corresponding positions of the new solution.

Step 4: delete those numbers of the substring from V and the remaining numbers of V form a subsequence.

Step 5: place the numbers of the subsequence into the unfilled positions of the new solution from left to right. Finally, the new solution is generated. The above procedures are shown in Fig. 8.4.

8.4.4 *Hybridizing ABC with a New Local Search*

In traditional local search, it begins with an initial solution as current solution and then generates a neighborhood solution based on the current one for exploiting a better solution. If the neighbor is better than the current one, the local search will designate the neighbor as the current solution and then repeat the procedure. Otherwise, it will stop. In order to improve the exploitation of proposed ABC, an

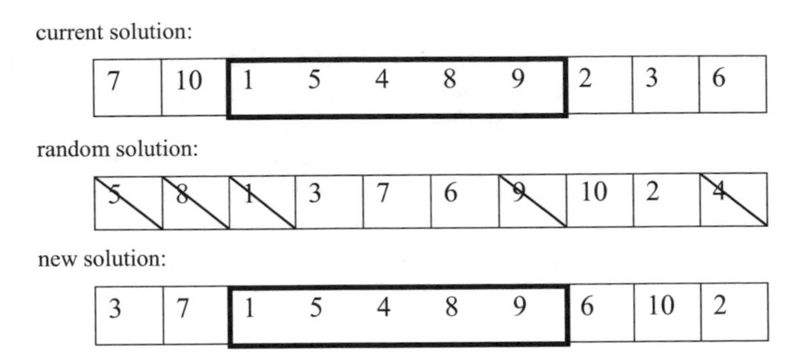

Fig. 8.4 Steps to generate a new solution where the N is assumed to be 10, $a = 3$ and $b = 7$

improved local search is introduced to conduct the evolution. It starts from the solution just searched by onlooker bee and then moves to a neighbor solution by an operator so as to generate an improved solution.

The procedure of proposed local search is inspired by simulate annealing algorithm, which is shown in Fig. 8.5. The steps of the operator used to produce new individual are shown as follows:

Step 1: randomly select a job from the current solution S and then insert it into a randomly selected position of S to generate a neighbor solution U. This procedure is shown in Fig. 8.6.

Step 2: calculate the fitness value of the new produced solution (fit_U) and the old one (fit_S) respectively.

Step 3: compare the magnitude of fit_S and fit_U. If fit_U is bigger than fit_S, U will be used to replace S. Otherwise, U is accepted with a probability P_U that depends on the current iteration size T. The probability P_U is evaluated by Eq. 8.13 as follows:

$$P_U = \exp\left(-T \times (\text{fit}_S - \text{fit}_U)\right) \tag{8.13}$$

8.5 Experiments and Interpretation of Results

In the experimental stage, the probability ($C_{\max} \leq X$) of proposed ABC schedules is compared with the probability ($C_{\max} \leq X$) of other well-known heuristic algorithms to demonstrate the efficiency of proposed ABC for current problem. NEH heuristic, a famous method, is proposed by Nawaz, Enscore, and Ham [20], which has proven its advantages for M-machine permutation flowshop scheduling problem. Robust schedule with the objective of probability ($C_{\max} \leq X$) for M-machine permutation flowshop problem was first presented by Liu et al. [19], in which the robustness of an improved genetic algorithm has been proved to be better than the robustness of NEH

The pseudo code of local search

For *i =1 to SN %SN denotes the population size%*

Generate a neighbor solution V_i *according to* X_i;

$Q = f(X_i) - f(V_i)$;

If $(Q < 0)$

$X_i \leftarrow V_i$;

Else

If $(Q = 0)$

Accept V_i *with probability of 0.5;*

Else

Accept V_i *with probability of* $\exp(-T \times Q)$; *%T is the current iteration size %*

End if

End if

End for

Fig. 8.5 The pseudo code of local search

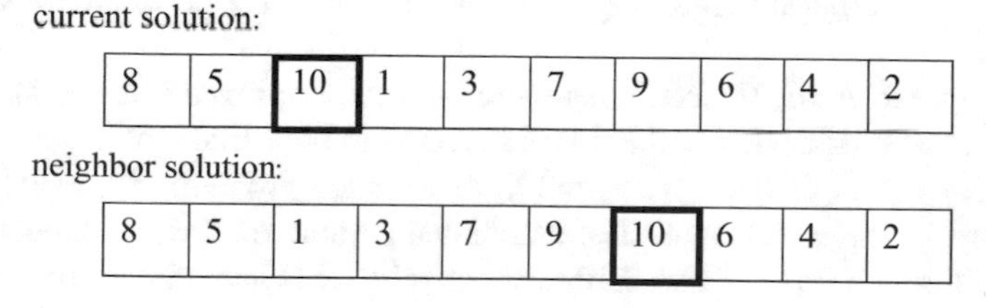

Fig. 8.6 The procedure of generating a neighbor solution. It is assumed that the random job is 10 and the random position is 7

heuristic in terms of the objective of probability ($C_{\max} \leq X$). Therefore, in order to certify the efficiency of our algorithm, the proposed ABC will be compared with the improved GA developed by Liu et al. [19] based on the probability ($C_{\max} \leq X$).

When calculating the probability ($C_{\max} \leq X$), the expected finish time is needed. In order to realize the comparison between proposed ABC and improved GA, some assumptions are used. The probability ($C_{\max} \leq X$) of NEH heuristic schedules will be assumed and the associated value z is calculated through Eqs. 8.7 and 8.10. The value z is utilized to evaluate the expected finish time of jobs by Eq. 8.8. Ultimately,

Table 8.1 Mean and variance of processing tine of jobs for Car 1 problem (11 × 5)

Car 1 problem (11 × 5)										
Jobs	Mean processing times at machine					Variance of processing time at machine				
	m1	m2	m3	m4	m5	m1	m2	m3	m4	m5
1	375	12	142	245	412	36	1	6	3	11
2	632	452	758	278	398	6	16	1	16	34
3	12	876	124	534	765	1	4	11	50	61
4	460	542	523	120	499	13	20	31	6	10
5	528	101	789	124	999	26	3	18	9	91
6	796	245	632	375	123	76	2	54	20	6
7	532	230	543	896	452	24	12	7	35	25
8	14	124	214	543	785	1	12	17	26	78
9	257	527	753	210	463	28	49	44	23	34
10	896	896	214	258	259	48	33	5	1	23
11	532	302	501	765	988	43	29	13	26	86

the expected finish time is applied to compute probability ($C_{\max} \leq X$) of proposed ABC schedules and the improved GA schedules, respectively.

Permutation flowshop problems Car 1 to Car 8 and Rec 1 to Rec 17 taken from operations research (OR) library are analyzed in this experiment. Since the processing times of jobs in those datasets in the OR library are deterministic, it is assumed that these deterministic times are the mean processing times of jobs. The variance of associated processing time is randomly selected from an interval of $[1, \mu_{ij}/9]$ which the μ_{ij} represents the mean processing time of job i on machine j. The mean and variance processing times of jobs for Car 1 problem are showed in Table 8.1.

The proposed ABC, the NEH heuristic, and the improved GA are all coded in Visual C for analysis. NEH heuristics is assumed to have three different probability ($C_{\max} \leq X$) values of 0.80, 0.85, and 0.90 for each problem. So, every problem corresponds to three instances. The associated z value of each probability is then calculated through Eqs. 8.7 and 8.10, respectively. NEH heuristic firstly solves each instance to obtain the mean and variance makespan of jobs and then evaluates the associated expected finish time by using Eq. 8.8. The expected finish time is denoted by X showed in Tables 8.2 and 8.3.

Each instance is then solved by proposed ABC and improved GA to get their respective mean and variance makespan. The expected finish time X of each instance obtained from NEH schedules is used to represent the expected finish time of proposed ABC schedules and improved GA schedules. Then the corresponding z value is computed by Eq. 8.8. Finally, the probability value of proposed ABC schedules and improved GA schedules for each instance is evaluated using Eqs. 8.7 and 8.10. At last, for better analysis of the comparison between proposed ABC and improved GA, proportion increase in probability ($C_{\max} \leq X$) and proportion decline in risk for each instance are introduced, which are shown in Eqs. 8.14 and 8.15.

Table 8.2 Results for Car problems

Problem	Size	NEH				Genetic algorithm				Artificial bee colony algorithm				Percent increase in probability in ABC over GA	Percent decrease in risk in ABC over GA
		Probability	φ(z)	$C_{\max}$	X	φ(z)	Probability	$C_{\max}$	X	φ(z)	Probability	$C_{\max}$	X		
Car 1	11 × 5	0.8	0.3	7038	7152.73	0.37	0.87	7050	7152.73	0.362	0.862	7040	7152.73	−2.16	−1.25
		0.85	0.35	7038	7179.78	0.417	0.917	7050	7179.78	0.413	0.913	7040	7179.78	−0.96	−0.68
		0.9	0.4	7038	7212.56	0.434	0.934	7059	7212.56	0.457	0.957	7040	7212.56	5.30	4.24
Car 2	13 × 4	0.8	0.3	7376	7534.52	0.5	1	7180	7534.52	0.5	1	7166	7534.52	0	0
		0.85	0.35	7376	7571.89	0.5	1	7176	7571.89	0.5	1	7166	7571.89	0	0
		0.9	0.4	7376	7617.19	0.5	1	7200	7617.19	0.5	1	7186	7617.19	0	0
Car 3	12 × 5	0.8	0.3	7399	7543.52	0.422	0.922	7403	7543.52	0.45	0.95	7312	7543.52	6.64	5.09
		0.85	0.35	7399	7577.59	0.425	0.925	7312	7577.59	0.472	0.972	7312	7577.59	11.06	8.9
		0.9	0.4	7399	7618.9	0.447	0.947	7401	7618.9	0.484	0.984	7312	7618.9	8.28	7.17
Car 4	14 × 4	0.8	0.3	8129	8270.02	0.5	1	8024	8270.02	0.5	1	8011	8270.02	0	0
		0.85	0.35	8129	8303.26	0.5	1	8004	8303.26	0.5	1	8014	8303.26	0	0
		0.9	0.4	8129	8343.56	0.5	1	8018	8343.56	0.5	1	8106	8343.56	0	0
Car 5	10 × 6	0.8	0.3	7835	7946.51	0.469	0.969	7758	7946.51	0.466	0.966	7727	7946.51	−0.64	−0.56
		0.85	0.35	7835	7972.8	0.47	0.97	7798	7972.8	0.479	0.979	7727	7972.8	1.91	1.73
		0.9	0.4	7835	8004.67	0.49	0.99	7758	8004.67	0.49	0.99	7727	8004.67	0	0
Car 6	8 × 9	0.8	0.3	8773	8893.55	0.5	1	8570	8893.55	0.5	1	8505	8893.55	0	0
		0.85	0.35	8773	8921.97	0.5	1	8570	8921.97	0.5	1	8505	8921.97	0	0
		0.9	0.4	8773	8956.43	0.5	1	8570	8956.43	0.5	1	8505	8956.43	0	0
Car 7	7 × 7	0.8	0.3	6590	6702.77	0.254	0.754	6590	6702.77	0.381	0.881	6590	6702.77	50	20.51
		0.85	0.35	6590	6729.35	0.3	0.8	6590	6729.35	0.43	0.93	6590	6729.35	43.33	22.8
		0.9	0.4	6590	6761.58	0.35	0.85	6590	6761.58	0.468	0.968	6590	6761.58	33.71	22.18
Car 8	8 × 8	0.8	0.3	8564	8694.23	0.5	1	8366	8694.23	0.5	1	8424	8694.23	0	0
		0.85	0.35	8564	8724.94	0.5	1	8366	8724.94	0.5	1	8420	8724.94	0	0
		0.9	0.4	8564	8762.16	0.5	1	8429	8762.16	0.5	1	8424	8762.16	0	0

Table 8.3 Results for Rec problems

Problem	Size	NEH				Genetic algorithm				Artificial bee colony algorithm				Percentage increase in probability in ABC over GA	Percent decrease in risk in ABC over GA
		Probability	$\varphi(z)$	C_{max}	X	$\varphi(z)$	Probability	C_{max}	X	$\varphi(z)$	Probability	C_{max}	X		
Rec 1	20 × 5	0.8	0.3	1320	1340.82	0.5	1	1271	1340.82	0.5	1	1279	1340.82	0	0
		0.85	0.35	1320	1345.72	0.5	1	1268	1345.72	0.5	1	1291	1345.72	0	0
		0.9	0.4	1320	1351.67	0.5	1	1288	1351.67	0.5	1	1293	1351.67	0	0
Rec 3	20 × 5	0.8	0.3	1116	1133.02	0.312	0.812	1115	1133.02	0.452	0.952	1111	1133.02	44.87	25.55
		0.85	0.35	1116	1137.03	0.397	0.897	1111	1137.03	0.475	0.975	1112	1137.03	19.65	14.86
		0.9	0.4	1116	1141.89	0.437	0.937	1111	1141.89	0.49	0.99	1113	1141.89	12.13	10.39
Rec 5	20 × 5	0.8	0.3	1296	1315.8	0.5	1	1255	1315.8	0.5	1	1261	1315.8	0	0
		0.85	0.35	1296	1320.47	0.5	1	1247	1320.47	0.5	1	1266	1320.47	0	0
		0.9	0.4	1296	1326.13	0.5	1	1268	1326.13	0.5	1	1277	1326.13	0	0
Rec 7	20 × 10	0.8	0.3	1626	1645.89	0.5	1	1584	1645.89	0.5	1	1584	1645.89	0	0
		0.85	0.35	1626	1650.58	0.5	1	1599	1650.58	0.5	1	1584	1650.58	0	0
		0.9	0.4	1626	1656.26	0.5	1	1601	1656.26	0.5	1	1596	1656.26	0	0
Rec 9	20 × 10	0.8	0.3	1583	1602.4	0.34	0.84	1580	1602.4	0.5	1	1557	1602.4	47.06	32
		0.85	0.35	1583	1606.98	0.448	0.948	1575	1606.98	0.5	1	1557	1606.98	11.61	10.4
		0.9	0.4	1583	1612.52	0.464	0.964	1574	1612.52	0.5	1	1547	1612.52	7.76	7.2

Rec 11	20 × 10	0.8	0.3	1550	1569.91	0.5	1	1502	1569.91	0.5	1	1512	1569.91	0	0
		0.85	0.35	1550	1574.6	0.5	1	1509	1574.6	0.5	1	1501	1574.6	0	0
		0.9	0.4	1550	1580.29	0.5	1	1491	1580.29	0.5	1	1513	1580.29	0	0
Rec 13	20 × 15	0.8	0.3	2002	2025.9	0.465	0.965	1981	2025.9	0.5	1	1956	2025.9	7.53	7
		0.85	0.35	2002	2031.54	0.5	1	1969	2031.54	0.5	1	1964	2031.54	0	0
		0.9	0.4	2002	2038.37	0.5	1	1979	2038.37	0.5	1	1966	2038.37	0	0
Rec 15	20 × 15	0.8	0.3	2025	2045.59	0.49	0.99	1986	2045.59	0.5	1	1983	2045.59	2.04	2
		0.85	0.35	2025	2050.45	0.49	0.99	1998	2050.45	0.5	1	1990	2050.45	2.04	2
		0.9	0.4	2025	2056.33	0.5	1	1997	2056.33	0.5	1	1981	2056.33	0	0
Rec 17	20 × 15	0.8	0.3	2019	2044.4	0.484	0.984	1992	2044.4	0.5	1	1971	2044.4	3.31	3.2
		0.85	0.35	2019	2050.39	0.5	1	1986	2050.39	0.5	1	1988	2050.39	0	0
		0.9	0.4	2019	2057.65	0.5	1	1992	2057.65	0.5	1	1994	2057.65	0	0

$$\text{Proportion}_{\text{increase\_probability}} = \frac{\varphi(z)_{\text{ABC}} - \varphi(z)_{\text{GA}}}{\varphi(z)_{\text{GA}}} \times 100\% \tag{8.14}$$

$$\text{Proportion}_{\text{decrease\_risk}} = \frac{\left(1 - \varphi(z)_{\text{GA}}\right) - \left(1 - \varphi(z)_{\text{ABC}}\right)}{1 - \varphi(z)_{\text{ABC}}} \times 100\% \tag{8.15}$$

8.5.1 Analysis of Car Problems

The results in Table 8.2 for Car problems show that the proposed novel ABC gets better robust schedules for most Car instances than the improved GA, only except the first and second instances of Car 1 and the first instance of Car 5. The results between the two algorithms for those three instances are very close. In addition, the probabilities for all instances obtained from proposed ABC are all larger than that of NEH heuristic. The probability of some instances for ABC schedules can reach 100%, which is the same as the improved GA schedules. The probability is due to the approximation used for calculating $\varphi(z)$. For ABC schedules, the maximum ratio rise in probability of 50% and maximum decline in risk of 22.81% are observed in Car 7 when compared with the improved GA, which are also better than the NEH heuristic. Figure 8.7a, b demonstrates that most results obtained from proposed ABC schedules can dramatically improve the probability of ensuring makespan under the expected finish time and decline the risk of exceeding a given limit time for Car problems.

8.5.2 Analysis of Rec Problems

The results for Rec problems are demonstrated in Table 8.3. The results of proposed ABC for Rec problems are all better than the improved GA and NEH heuristic. The probabilities for most Rec instances except Rec 3 obtained from proposed ABC schedules reach 100%. When compared with improved GA schedules, proposed ABC schedules get minimum ratio rise in probability of 2.04% with 2% decline in risk in Rec 15. However, the maximum proportion increase in probability of 47.06% is observed in Rec 9 with maximum fall in risk of 32%. Figure 8.8a, b indicates that for Rec problems, most of results obtained from proposed ABC schedules can greatly rise the likelihood of making sure that makespan not surpass expected finish time and fall the risk of exceeding makespan from a limited value.

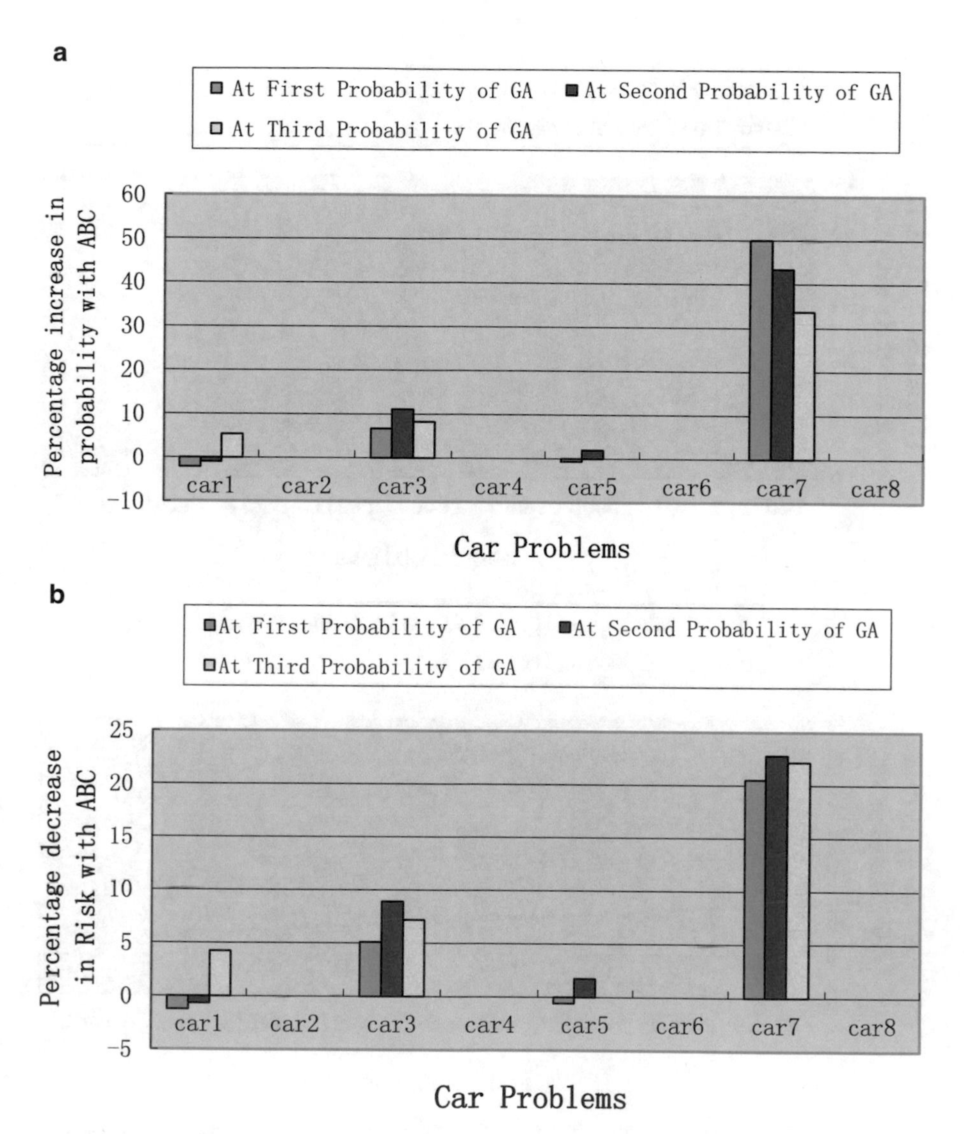

Fig. 8.7 (**a**) Comparison of probabilities on Car problems. (**b**) Comparison of risk on Car problems

8.5.3 Analysis of Overall Results

The average proportion rise in probability and mean proportion decline in risk for both Car and Rec problems are demonstrated in Fig. 8.9. When compared with the improved GA, proposed ABC schedules get over 14% improvement in

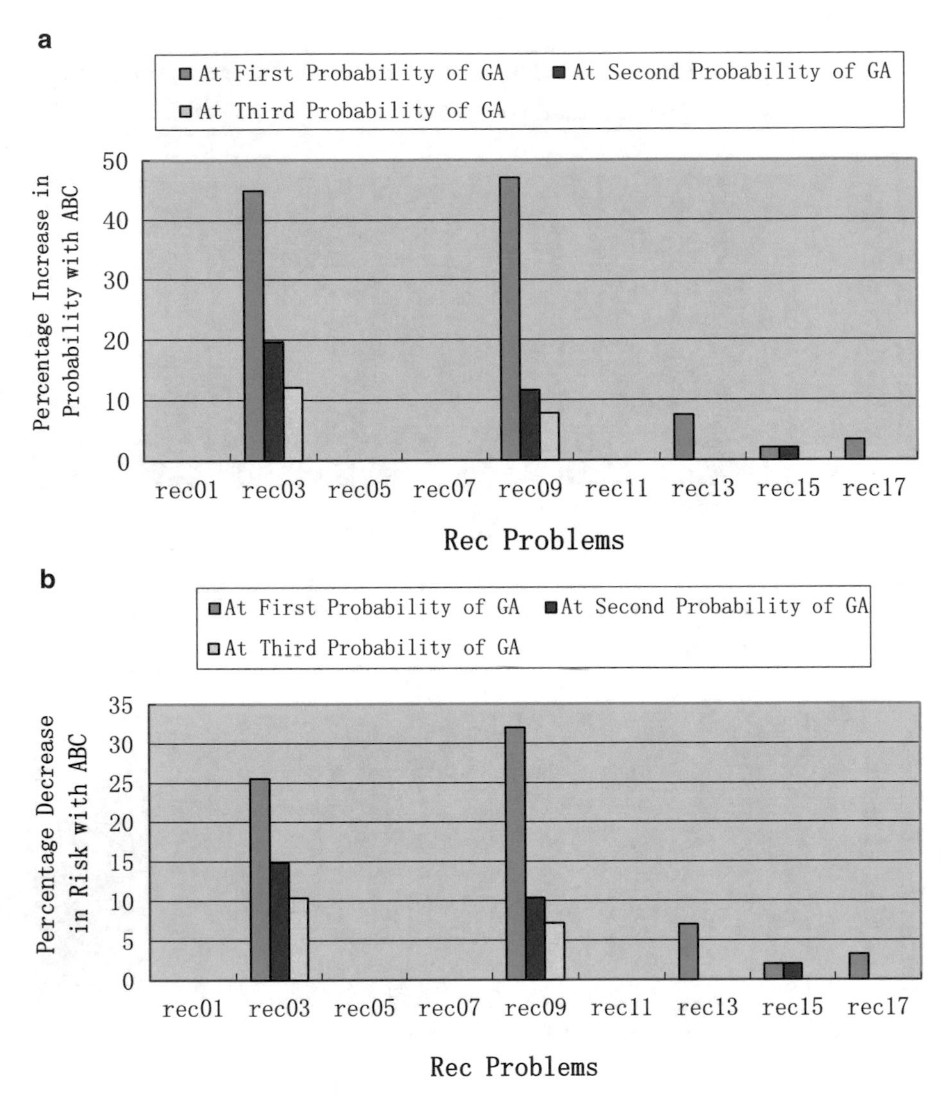

Fig. 8.8 (**a**) Comparison of probabilities on Rec problems. (**b**) Comparison of risk on Rec problems

probability of making makespan under an expected finish time for Car problems and over 15% for Rec problems. Similarly, proposed ABC schedules obtain more than 8% decline in risk of surpassing makespan over an expected limit time for Car problems and over 11% for Rec problems. The overall results show that the schedules obtained from proposed ABC can give a higher probability and a higher ratio decrease in risk when compared with the schedules generated by the improved GA for both Car and Rec problems. From what has been discussed

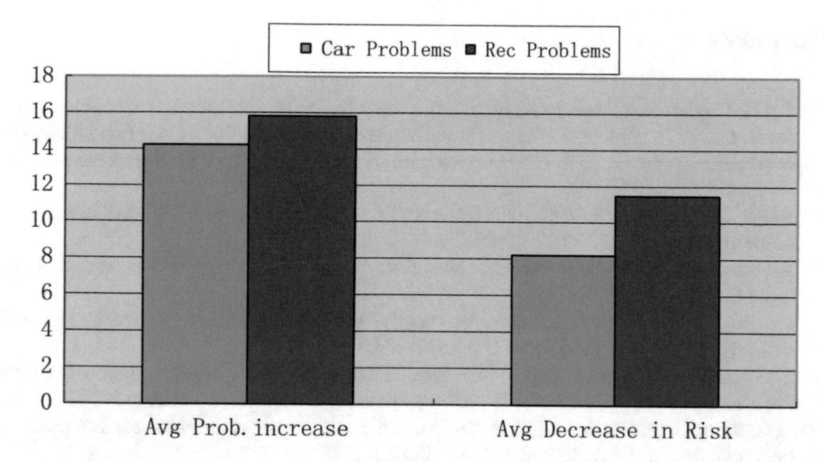

Fig. 8.9 Average percentage increase in probability and average percentage decrease in risk by ABC schedules

above, the proposed ABC demonstrates its excellent performance on the M-machine permutation flowshop problem and it is able to produce more robust schedules than other existed algorithms.

8.6 Conclusion

An M-machine permutation flowshop problem considering the uncertainty of the processing time of jobs is studied in this chapter. The purpose is to find a more robust schedule for this problem, which can give a higher probability of guaranteeing the makespan less than the expected finish time. According to the central limit theorem, the distribution of uncertain processing time of jobs is normally distributed. A novel artificial bee colony algorithm with an improved local search which can enhance the exploration and exploitation is proposed. The performance of proposed ABC is tested on the well-known Car and Rec permutation problems taken from the OR library and compared with an improved GA and NEH heuristic. The experiment results indicate that the schedules formed by proposed ABC are more robust than that of improved GA for most instances and better than NEH heuristic for all instances.

In this chapter, the considered problem is simplified that all jobs have the same expected finish time. But, the expected finish time of jobs might not be fixed in actual flowshop problems. So, finding robust schedules for permutation flowshop problems with different expected finish time is needed to be studied in future.

References

1. Aytug H, Lawley MA, McKay K, Mohan S, Uzsoy R (2005) Executing production schedules in the face of uncertainties: a review and some future directions. Eur J Oper Res 161:86–110
2. Pierreval H, Durieux-Paris S (2007) Robust simulation with a base environmental scenario. Eur J Oper Res 182:783–793
3. Goren S, Sabuncuoglu I (2008) Robustness and stability measures for scheduling: single-machine environment. IIE Trans 40:66–83
4. Li Z, Ierapetritou M (2008) Process scheduling under uncertainty: review and challenges. Comput Chem Eng 32:715–727
5. Herroeleny W, Leus R (2004) Robust and reactive project scheduling: a review and classification of procedures. Int J Prod Res 42(8):1599–1620
6. Lan S, Clarke J, Barnhart C (2006) Planning for robust airline operations optimizing aircraft routings and flight departure times to minimize passenger disruptions. Transp Sci 40(1):15–28
7. Tang L, Wang X (2008) A predictive reactive scheduling method for color-coating production in steel industry. Int J Adv Manuf Technol 35:633–645
8. Feiza G, Henri P, Hajri-Gabouj S (2010) Analysis of robustness in proactive scheduling: a graphical approach. Comput Ind Eng 58:193–198
9. Sabuncuoglu I, Goren S (2009) Hedging production schedules against uncertainty in manufacturing environment with a review of robustness and stability research. Int J Comput Integr Manuf 22(2):138–157
10. Kouvelis P, Daniels RL, Vairaktarakis G (2000) Robust scheduling of a two-machine flowshop with uncertain processing times. IIE Trans 32(5):421–432
11. Kuo CY, Lin FJ (2002) Relative robustness for single-machine scheduling problem with processing time uncertainty. J Chin Inst Ind Eng 19(5):59–67
12. Kouvelis P, Yu G (1997) Robust discrete optimization and its applications. Kluwer Academic Publishers, Boston
13. Daniels RL, Kouvelis P (1995) Robust scheduling to hedge against processing time uncertainty in single-stage production. Manag Sci 41:363–376
14. Kaperski A (2005) Minimizing maximal regret in the single machine sequencing problem with maximum lateness criterion. Oper Res Lett 33:431–436
15. Yang J, Yu G (2002) On the robust single machine scheduling problem. J Comb Optim 6:17–33
16. Liu L, Gu HY, Xi YG (2007) Robust and stable scheduling of a single machine with random machine breakdowns. Int J Adv Manuf Technol 31:645–654
17. Daniels RL, Carrillo JE (1997) β-Robust scheduling for single machine systems with uncertain processing times. IIE Trans 29:977–985
18. Wu CW, Brown KN, Beck JC (2009) Scheduling with uncertain durations: modeling β-robust scheduling with constraints. Comput Oper Res 36:2348–2356
19. Liu Q, Ullah S, Zhang C (2011) An improved genetic algorithm for robust permutation flowshop scheduling. Int J Adv Manuf Technol 56:345–354
20. Nawaz M, Enscore E, Ham I (1983) A heuristic algorithm for the m-machine, n-machine flow shop sequencing problem. Omega 11:91–95
21. Teixidor AB (2006) Proactive management of uncertainty to improve scheduling robustness in process industries. PhD thesis, Universitat Politμecnica de Catalunya Barcelona
22. Hapke M, Jaskievicz A, Slowinski R (1999) Fuzzy multimode resource-constrained project scheduling with multiple objectives. In: Weglarz J (ed) Project scheduling recent models, algorithms and applications. Kluwer Academic Publishers, Dordrecht, pp 355–382
23. Allen E (2007) Modeling with Itô stochastic differential equation. Springer, Dordrecht
24. Hayter AJ (1996) Probability and statistics for engineers and scientists. PWB Publishing, Duxbury
25. Karaboga D, Basturk B (2007) A powerful and efficient algorithm for numerical function optimization: artificial bee colony (ABC) algorithm. J Glob Optim 39(3):459–471
26. Karaboga D, Basturk B (2008) On the performance of artificial bee colony (ABC) algorithm. Appl Soft Comput 8(1):687–697

Chapter 9
Semantic Image Retrieval Using Point-Set Topology and the Ant Sleeping Model

Deepak Karunakaran and Shrisha Rao

9.1 Introduction

Collections of digital multimedia are growing at extremely high rates, and their organization and management are crucial. The proliferation of multimedia devices like cameras and video recorders has necessitated a focus on new ways to handle these data through search engines, digital libraries, etc. Content-based image retrieval (CBIR) is an important paradigm to address this issue of efficient search and retrieval of images from large image databases. It deals with challenges like feature extraction, high-dimensional indexing and storage, semantic analysis, visualization, multimodal retrieval, etc. [16, 70].

Feature extraction [22] is typically the extraction of feature vectors from images which could be used to determine the logical distance between images as a measurement of similarity [62]. These feature vectors are generally of very high dimensions, making their handling computationally expensive for the retrieval algorithms. On the one hand, there is the well-known problem of "the curse of dimensionality"—an increase in the dimensionality of the features does not always imply better classification, but may only cause difficulties in feature selection [21]. And on the other hand is the problem of storing and retrieving these features with the increasing scales of very large image databases [2]. Semantic analysis [38] is used to address the challenge of connecting image features to the semantic meaning in the image. The conventional content-based image retrieval systems

D. Karunakaran (✉)
Victoria University of Wellington, Wellington, New Zealand
e-mail: deepakk@ieee.org

S. Rao
International Institute of Information Technology – Bangalore, Bengaluru, India
e-mail: shrao@ieee.org

X. Li, K.-C. Wong (eds.), *Natural Computing for Unsupervised Learning*,
Unsupervised and Semi-Supervised Learning,
https://doi.org/10.1007/978-3-319-98566-4_9

perform poorly due to the "semantic gap." To deal with semantic gap, relevance feedback [52], exploring ontological models for image annotation [53], and using multiple information sources, viz., textual information, are some of the approaches.

We develop the premise of this chapter by outlining the challenges with image indexing and semantic retrieval. The problem of image search has evolved over the years, and an exponentially increasing scale is one of the major challenges. In other words, "How to build an efficient index to scale up an image retrieval system?" [70]. Another major challenge is to apply user feedback to improve on image search engine performance [70]. This, in particular, is a highly complex problem requiring detection of deep underlying patterns. For such problems nature-inspired algorithms, e.g., [17, 18, 29, 37], are typically more effective.

9.1.1 Multidimensional Indexing

Feature vectors (e.g., MPEG-7 descriptors) representing images form the objects of an image database which is queried for similarity search in a CBIR system. These feature vectors are typically of very high dimensions. Performing a similarity search is thus computationally expensive. These features need to be stored on a hard disk or similar device. The computational speeds have increased exponentially over recent decades having followed Moore's law, but the speeds of storage devices are lagging, owing to the presence of mechanical components [50]. This makes the similarity search, e.g., the k-nearest neighbor search in large, high-dimensional datasets, very challenging, with the search time dominated by disk access times. With issues in semiconductor memory [55] delaying its widespread adoption at large scales, this is a pertinent problem.

Indexing a database results in its reorganization to enable fast search for required nearest neighbors. A number of index structures have been proposed for handling the search and retrieval of multidimensional data [10]. Most early approaches (viz., R-tree [23], SS-tree [66], etc.) divide the data space according to the distribution of data objects inserted or loaded into the tree. When the dimensionality of the feature space increases, these index structures fare poorly [65], seen as a result of *the curse of dimensionality* [4].

Recent approaches to nearest neighbor search include vector approximation techniques [19] which work over a compressed representation of the dataset; vector quantization techniques [11, 12, 61] are an example. Another approach toward ameliorating the curse of dimensionality is to map the higher-dimensional space to a lower dimension which is known as dimensionality reduction [72]. This enables the indexing structures to work better. Some other classes of methods which come under its purview are variable-subset selection [6], multidimensional scaling [60, 64], and hashing [9, 39, 67].

9.1.2 Semantic Retrieval with Relevance Feedback

Relevance feedback [52] is a mechanism to narrow the gap between high-level concepts and low-level features and to take into account the subjectivity of the user's perception—for example, say a user is interested only in a dingo and not in any other species of dog; how then to map low-level features with this high-level concept? Similarly, user perception of the images may be different, e.g. [52], one user is interested more in the color feature and another in the texture feature of the same image. The main idea of relevance feedback is to apply supervised learning from the positive and negative examples given by the user toward improving the information retrieval system [56]. Relevance feedback is thus used to reduce this *semantic gap*.

Relevance feedback (RF) could be generalized into "the weighing approach" and "the probability approach" [56]. Multidimensional indexing with either of these approaches suffers from the curse of dimensionality [56]. The former associates higher weights to important dimensions (importance decided by the relevance feedback) and the latter use methods, e.g., Bayesian learning, to minimize the retrieval error by updating the probability distribution of images in the database through user feedback.

Another categorization of the relevance feedback is done with a classification problem, a learning problem, and a distance optimization problem [30]. In a classification problem, the database of images is classified into relevant and irrelevant images, viz., using support vector machines (SVM) [59], Bayesian modeling [42], etc. In the case of a learning problem, the samples fed back by the user are used to train a model for retrieval, e.g., integrating reinforcement learning into RF [69], PicSOM [28], Bayesian Framework-based RF [15], and decision tree learning [40]. RF as a distance optimization problem solves for the parameters which weigh the features and transform the feature space to obtain better results, e.g., using non-parametric discriminant analysis [58]. In this chapter, we develop relevance feedback approaches which could be categorized into "the weighing approach" with the important distinction that we do not increase the weights of particular dimensions from user feedback. Instead we apply the relevance feedback over the efficiently indexed "intra-cluster similarities" and identify those parts of clusters that are most relevant to the user. Furthermore, our method is closer to the classifier problem; we cluster our image database and then index it for faster retrieval when relevance feedback is provided.

9.1.3 Ant-Based Systems

Nature-inspired computing deals with formulating computing systems modeled on population(s) of autonomous entities which are most effective when the problem characteristics include high complexity, dynamic environments, complex underlying patterns, and decentralized properties [36]. Ant-based systems have shown success in pattern analysis context, particularly clustering and topographic mapping [24, 44] which are inherently complex. In particular, ant colony systems by Dorigo et al. [17, 18] have been very popular for optimization problems. Some examples of other prominent works in pattern recognition and analysis are [35, 41, 47, 54]. Based on Maslow's hierarchy of needs theory, [68] develop an ant sleeping model (ASM) for clustering. In ASM, an artificial ant has two states, *active* and *sleep*. Ants in the active state move around a two-dimensional grid till they find similar ants in their neighborhood and change their state to *sleep*. The groups of sleeping ants form clusters, clusters of similar images in this case. In this chapter, we propose an algorithm based on ASM for relevance feedback and evaluate its performance *vis-à-vis* semantic retrieval accuracy and the retrieval time efficiency from storage.

In general, nature-inspired computing systems suffer from a lack of good mathematical foundations leading to misunderstanding and criticisms [8, 48]. Due to lack of mathematical frameworks, these algorithms are difficult to analyze. This is one of the reasons that these algorithms are yet to see a wider acceptance for industrial applications, barring some exceptions, though having shown good potential. To this end, we endeavor to develop a sound mathematical framework for our proposed ant-based system in the context of semantic image retrieval. We develop theoretical constructs using point-set topology to propose a generalized mathematical model for semantic analysis and use that to explain our ant-based relevance feedback system. We model the semantic concepts and images related to them as topological spaces and explain the relevance feedback mechanism using their mathematical properties. Furthermore, a theoretical and empirical comparison is done with PicSOM [34] which is another nature-inspired algorithm based on self-organizing maps for content-based image retrieval using relevance feedback.

This book chapter is organized as follows:

- In the next section, a general topological model (Sect. 9.2) to represent semantic retrieval of images from a database using relevance feedback is presented. The point-set topology of the semantic concepts and the images is used to explain a general relevance feedback system.
- We propose an efficient algorithm for multidimensional indexing of large image databases based on the ant sleeping model in Sect. 9.3.
- We then extend the theory to explain our semantic retrieval method based on the ASM (Sect. 9.3.3). We discuss precision and recall as measures of accuracy using the proposed model. Furthermore, using our model we construct a theorem to compare two nearest neighbor search-based algorithms which use relevance feedback. We extend the theorem to show how a compared algorithm can be more efficient for storing and retrieving images from a database.
- In Sect. 9.4, the experiments and their results are presented which show that our technique is computationally efficient when compared to the existing state-of-the-art indexing methods. We show that the average response time for retrieval of relevant images from a storage device is lower using the indexing technique based on ASM, for large image databases.

- In the last section, we compare our ASM approach, both theoretically and experimentally against PicSOM (Sect. 9.4.4). In our experiments using relevance feedbacks, the ASM indexing technique shows significant improvement in the retrieval performance over PicSOM.

9.2 General Topological Model for Semantic Image Retrieval

The richness of user semantics is a challenge for retrieval systems to associate accurate images to a user query. As an example, consider a scenario, where a user is interested in categorizing images of sports based on the equipment needed. Assume that we have an image database for sports. Now, a semantic concept of a "ball" will result in images from the category of ball sports, e.g., tennis, football, cricket, squash, golf, hockey, polo, baseball, etc. Similarly the semantic concept of "racket" will result in images from "racquet sports" like tennis, squash, badminton, etc. But a combination of the two semantic concepts as "racket and ball" will considerably restrict images to sports like tennis, squash, table tennis, etc. Similarly "racket without net" would restrict the images to say just those of squash. Now if we introduce the semantic concept of "Olympics," then the scope of the images related to each of these combinations is significantly altered by the introduction of just one concept.

The simple example above shows how the combination of different semantic concepts map to the different sets of images in the database. The mathematical domain of point-set topology provides us with theoretical constructs which are useful for such a representation. The images (features) and the semantics are represented as point sets. We use the set-theoretic constructs to relate the sets of semantic concepts to those (clusters) of images and develop a model based on general topology toward semantic analysis of the complex mapping of images to the semantic concepts.

General topology [45] or point-set topology deals with the fundamentals of set-theoretic definitions and constructions used in topology. Topology gives us the ability to classify and count objects using qualitative or approximate information as opposed to exact values [3]. For example, Nash equilibria deal with qualitative properties of interactions in game theory (applied in economics); the existence of Nash equilibria is proved using Brouwer's fixed point theorem in topology [46]. Similarly topology is employed in condensed-matter physics, cosmology, and study of many chaotic interactions to handle qualitative information to give provable resulting behavior [3].

The concept of topology emerged from the study of the real line, Euclidean space and the continuous functions on these spaces. A topological space is a set of points and the set of neighborhoods for each of the points, satisfying a set of axioms on these points and the neighborhoods. Topological space allows for the definition of concepts such as continuity, connectedness, and compactness which are fundamental in point-set topology. Consider a mapping from one topological space to another; if for each point, each neighborhood of the mapping of this point contains a neighborhood in the other topological space, then the mapping is continuous at the point. A connected space is a topological space that cannot be represented as the union of two or more disjoint non-empty open sets. Compactness generalizes the notion of closed and bounded subsets of the real line to topological spaces. Refer to [3, 45] for in-depth understanding.

These concepts enable us to model the semantic concept(s) and define their characteristics using general topology. We use this mathematical construct to propose a general model of semantic retrieval of images using relevance feedback. We construct topological spaces from the sets of semantic concepts and images and discuss their mathematical properties under the purview of relevance feedback. In Sect. 9.3.3 we use this general model to describe our proposed method. We further discuss the precision and recall metrics used for evaluation of the image retrieval systems using our model. We also construct a theorem to compare any two algorithms for image retrieval using relevance feedback and extend the same as previously indicated.

The general definition of information by Floridi [20] states that σ is an information, understood as "semantic content" if and only if it consists of one or more data; the data in σ is well-formed; and the well-formed data are meaningful. Under the purview of an image retrieval system, we consider the low-level features of an image as the data which is well-formed, and the semantic concept(s) is its meaning. As an example, consider an image database created by images randomly obtained from the Internet. If an arbitrary user wants to retrieve all the images which have a "hill with a sunset," "hill" and "sunset" are two semantic concepts, and the low-level features of the images represent its data.

We denote $\mathfrak{I}$ as the set of images in a database which is queried by a user to retrieve images matching one or more semantic concepts. Let $\mathfrak{S}$ be the set of all possible semantic concepts associated with any of the images in the database of images, $\mathfrak{I}$. We define $T_{\mathfrak{I}}$ as the topology on the set $\mathfrak{I}$ and $(\mathfrak{I}, T_{\mathfrak{I}})$ as the corresponding topological space [45]. $(\mathfrak{I}, T_{\mathfrak{I}})$ is a *discrete topology* [45]. Similarly, $T_{\mathfrak{S}}$ is a topology defined on the set $\mathfrak{S}$, and $(\mathfrak{S}, T_{\mathfrak{S}})$ is the corresponding topological space. $T_{\mathfrak{S}}$ is a discrete topology. $\mathfrak{O}$ (oracle) maps every image to the set of semantic concepts associated with it using the function (f_{orc}), i.e., $f_{\text{orc}} : (\mathfrak{I}, T) \rightarrow (\mathfrak{S}, T_{\mathfrak{S}})$. Since $T_{\mathfrak{I}}$ is a discrete topology, the mapping f_{orc} is a continuous mapping.

Example Consider the image database $\mathfrak{I}$ to consist of just 12 images with the corresponding semantic content as described in the Table 9.1.

The discrete topological space $(\mathfrak{I}, T_{\mathfrak{I}})$ includes 2^{12}= 4096 open sets. For this example, the set $\mathfrak{S} = \{\mathfrak{s}, \mathfrak{h}, \mathfrak{c}, \mathfrak{p}\}$.

The discrete topology over the set $\mathfrak{S}$ is

$$T_{\mathfrak{S}} = \{\mathfrak{s}, \mathfrak{h}, \mathfrak{c}, \mathfrak{p}, \{\mathfrak{s}, \mathfrak{h}\}, \{\mathfrak{s}, \mathfrak{c}\}, \{\mathfrak{s}, \mathfrak{p}\}, \{\mathfrak{h}, \mathfrak{c}\}, \{\mathfrak{h}, \mathfrak{p}\}, \{\mathfrak{c}, \mathfrak{p}\}, \{\mathfrak{s}, \mathfrak{h}, \mathfrak{c}\}, \{\mathfrak{s}, \mathfrak{h}, \mathfrak{p}\}, \\ \{\mathfrak{s}, \mathfrak{c}, \mathfrak{p}\}, \{\mathfrak{h}, \mathfrak{c}, \mathfrak{p}\}, \{\mathfrak{s}, \mathfrak{h}, \mathfrak{c}, \mathfrak{p}\}, \phi\}$$

Table 9.1 Images and their semantic content

Semantic content	Description
$\mathfrak{s}$	Sun
$\mathfrak{h}$	Hill
$\mathfrak{c}$	Car
$\mathfrak{p}$	Peacock
Image	Description
s_1, s_2, s_3, s_4	Images of sun
k_1, k_2	Images of the sun on the hill side
c_1, c_2	Images of car(s)
z_1, z_2	Images of the sun on a hill side along with a car(s)
h_1, h_2	Images of hill(s)
p_1, p_2	Images of peacock(s)

The following are the examples of mapping done by $\mathfrak{O}$.

$$f_{\text{orc}}(\{sun\_hill\}) = \{\mathfrak{s}, \mathfrak{sh}, \mathfrak{h}\},$$

$$f_{\text{orc}}(\{sun\_hill, peacock\}) = \{\mathfrak{s}, \mathfrak{sh}, \mathfrak{h}, \mathfrak{p}\}$$

9.2.1 Formal Model Description

We model the image retrieval through relevance feedback using the mathematical constructs developed. When the user queries an image database, he uses a query image say i_q which actually represents a set of semantic concepts the user is interested in. Let us represent the set of these semantic concepts by S_q. We assume that $S_q \in (\mathfrak{S}, T_{\mathfrak{S}})$. Otherwise no relevant image should be retrieved from $\mathfrak{I}$.

In a typical relevance feedback system [71], a user queries the system with an input image, then selects the matching images from a set of resulting images, and continues doing so till the desired output is achieved. We develop our model further toward relevance feedback as follows:

- The user queries the database using an image or a keyword. The system produces an initial set of images as the result. Let $\mathscr{A}$ be an algorithm which takes $\mathfrak{I}$ and i_q as input and produces I_k^* as the set of output images such that $I_k^* \in \mathfrak{I}$.
- The user selects a set of images which match his query and gives back to the system as new input. Let I_{rk} be the set of the images which are given to the system as input. Again the algorithm $\mathscr{A}$ takes I_{rk} as input and produces I_{k+1}^* as the output. $I_{rk} \subseteq I_n^*$, $I_n^* \in \mathfrak{I}$ is the final result after n relevance feedback.
- The previous step is repeated till the user gets the desired output.

We represent the model for relevance feedback below:

$$\xrightarrow[init.query]{i_q} \mathscr{A}^0 \xrightarrow[init.result]{I_0^*} \text{User} \xrightarrow[rel.fb.]{I_{r1}} \mathscr{A}^1 \xrightarrow{I_1^*} \text{User} \xrightarrow[rel.fb.]{I_{r2}}$$
$$\ldots \quad \mathscr{A}^n \xrightarrow{I_n^*} \text{User}$$

If the algorithm $\mathscr{A}$ is 100% accurate, then $f_{\text{orc}}(I_n^*) = S_q$. In other words, after n relevance feedbacks from the user, she is able to retrieve all the matching images from the image database $\mathfrak{I}$. The superscripts of $\mathscr{A}^m$ denote the algorithm iteration for the mth relevance feedback. Also, $\mathscr{A}^0, \mathscr{A}^1, \ldots \mathscr{A}^n$ should *improve* the result such that $I_1^* \subseteq I_2^* \subseteq I_3^* \ldots \subseteq I_n^*$. In a relevance feedback system, for any two $\mathscr{A}^m$ and $\mathscr{A}^{m-1}$, the user feedbacks from the $m-1$ relevance feedback are used as inputs for $\mathscr{A}^m$.

The relevance feedback from the user is I_{rk}, where $k = 1 \ldots n$. Let us define $\mathscr{D}$ such that $f_{\text{orc}}(S_q) = \mathscr{D}$. In the next section, we extend the mathematical constructs just developed to explain our relevance feedback system based on ASM.

9.3 Indexing Using Ant Sleeping Model

We give a brief overview of the A^4C algorithm [13, 68] and then explain our method to use it for indexing. Unless stated otherwise, we use the same notation as in [13] to ensure clarity in understanding. Refer Table 9.2 for the symbols used and their definitions.

9.3.1 A^4C Clustering

A^4C clustering [13, 68] is a non-deterministic algorithm based on the ant sleeping model (ASM). In ASM, each ant is characterized by two states: *sleep* and *active*. An ant represents the feature descriptors (e.g., color features, shape features, etc.) of one image. Chen et al. and Xu et al. [13, 68] use the term *agent* for this. The ant's state is controlled by the fitness function which is defined over its local information. Ants move dynamically to form different subgroups, and the data objects they represent get clustered thereby. In our experiments these data objects correspond to the image descriptors.

We initialize the ants on a rectangular grid where each ant has eight neighbors. In order to determine the state of an ant, we use a fitness function which uses the neighborhood information of an ant over the rectangular grid. Initially all the ants are *active*. For an ant, if the other ants in the neighborhood show similarity with each other, it is assigned a high fitness value. A higher fitness value increases the probability of an ant to reach the state of *sleep*. This results in formation of clusters of similar ants across the grid. It is possible to obtain clusters which are similar to each other but spatially far apart on the grid.

Table 9.2 Notation

Symbol	Definition
ϖ_i	An *ant* which is a vector of image descriptors
grid	The toroidal grid to place the ϖ on *cells* for clustering
cell	A position in the grid
δ_i	(x, y) location of ϖ_i in the grid
n	Total number of ants in the grid
$f(\varpi_i)$	The *fitness* of an ant, $f(\varpi_i) \in [0, 1]$
$p(\varpi_i)$	The active probability of the ϖ_i
s_x,s_y	$s_x \in Z^+$ and $s_y \in Z^+$ are the vision limits for the ϖ in the horizontal and the vertical directions respectively
α	The average distance (Euclidean) between ϖ_i and other ants
$N(\varpi_i)$	Denotes the neighbor cells of ϖ_i
$\mathfrak{C}$	The new arrangement of ants (ϖ) in the grid after clustering
ϖ_s	The search ant which is the representation of the query image
η	($\eta \ll n, \eta \in \mathbb{N}$) the number of random initialization locations for ϖ_s
ϕ	($\phi \in \mathbb{N}$) the top ϕ fitness values considered relevant
Ω_i	An array of fitness values of ϖ_i at the locations of all other ants
κ	($\kappa \in \mathbb{N}$) the number of images shown during relevance feedback
Δ	The output set of similar images
N^{RF}	The number of relevance feedbacks used

We extract the image descriptors of each image in our database, and the resulting vectors are represented by the ants (ϖ). These ants are randomly placed on the CA *grid*. For every iteration of the A^{4}C algorithm, the *fitness* is calculated using Eq. (9.1) [13, 68] for each ant.

$$f(\varpi_i) = max \left\{ 0, \frac{1}{S} \sum_{\varpi_j} \left(1 - \frac{d(\varpi_i, \varpi_j)}{\alpha} \right) \right\} \tag{9.1}$$

where

$$\varpi_j \in N(\varpi_i) \ , \ S = \frac{1}{(2s_x + 1)\left(2s_y + 1\right)}$$

and

$$\alpha = \frac{1}{n \times (n-1)} \sum_i \sum_j d(\varpi_i, \varpi_j) \tag{9.2}$$

We have used $s_x = s_y = 1$ implying each ϖ has eight neighbors. The $fitness$ is used to calculate the active probability p (Eq. 9.3); if the value of p is lower than a threshold, then the ant is in sleep state; it is active otherwise.

$$p(\varpi_i) = \frac{\beta^\lambda}{\beta^\lambda + f(\varpi_i)^\lambda} \tag{9.3}$$

Here $\beta \in R^+$ is the threshold of the ant's fitness and $\lambda \in \mathbb{R}^+$ is the active pressure of the agents. We use $\lambda = 2$. We use a threshold for the probability value above which a ϖ is in "sleep," i.e., the ϖ is clustered with similar ants with a high probability and need not move further.

If the ϖ is active, then it moves randomly to a new location δ. The new location δ belongs to $N(\varpi_i)$. We decide the probability threshold value based on the results from experiments, as seen in Sect. 9.4.

After a sufficient number of such iterations, similar ants (ϖ) get clustered together, and most of them attain the state of sleep to form the *clustered space*, $\mathfrak{C}$. Refer to [13, 68] for more details of the algorithm.

9.3.2 Indexing and Retrieval Algorithms

We explain our method using the two algorithms: Algorithm 1 for generating the index which is a one-time and offline process and the Algorithm 2 to retrieve the relevant images using relevance feedback over the generated index.

In Algorithm 1, the methods from the previous section are used to generate an index of images which basically summarize images using the information from clusters. As mentioned in the previous section, similar ants can get grouped into separate clusters. The index generated spans across such clusters and captures the information about each ant's similarity across the locations on grid. In the Algorithm 2, this information is used in relevance feedback by making it computationally feasible to retrieve semantically similar images online. Moreover the index enables us to store the images in a more efficient way on the physical storage device with faster access times, especially in the case of relevance feedback needing multiple fast retrievals. We analyze this feature theoretically in the Corollary 1 and experimentally in Sect. 9.4.1.

9.3.2.1 Clustering and Index Generation

Once we obtain the clustered space $\mathfrak{C}$, we apply Algorithm 1 to generate the index needed for retrieval. For every ϖ_i in the clustered space, we create the set Ω_i, which contains those ants from the clustered space which are *most* similar to ϖ_i. The similarity is identified by the fitness values. This set created for all ants forms the **index**.

Table 9.3 Representative generated index: $\Omega_i \forall i \in [1, n]$

Number	Ant	Location	Fitness@Location
1	ϖ_1	Location of ϖ_{42}	0.1966
2	ϖ_1	Location of ϖ_1	0.1785
3	ϖ_1	Location of ϖ_{28}	0.1577
4	ϖ_1	Location of ϖ_{300}	0.0985
.	.	.	.
.	.	.	.
.	.	.	.
n	ϖ_1	Location of ϖ_x	0.0067
$n+1$	ϖ_2	Location of ϖ_{100}	0.18567
.	.	.	.
$2n$	ϖ_2	Location of ϖ_x	0.0089
.	.	.	.
.	.	.	.
$(n-1)n+1$	ϖ_n	.	.
.	.	.	.
n^2	ϖ_n	.	.

Algorithm 1: Index generation

```
Input: 𝔈
Output: Ω_1, Ω_2, ..., Ω_n
for i ← 1 : n do
    for j ← 1 : n do
        Store f(ϖ_i) calculated at δ_j and position δ_j into Ω_i ;
    end
    Sort Ω_i in descending order of fitness values;
    Retain top-k elements in Ω_i;
end
```

Table 9.3 shows an arbitrary example of the index generated from Algorithm 1. The "Location" column lists those locations for an ant (ϖ) in the column "Ant" which have highest fitness in the clustered space. So in other words, if we take the example of ϖ_1, the locations of the ants ϖ_{42}, ϖ_{28}, and ϖ_{300} are exactly the locations where ϖ_1 has high fitness value, and these ants are therefore *similar* to ϖ_1. Line 5 of Algorithm 1 is used to create the rows of Table 9.3, where ϖ_i, δ_j, and $f(\varpi_i)$ form the columns Ant, Location, and Fitness@Location, respectively.

Figure 9.2 shows a small portion of the *grid* after clustering. We use 20,000 random images from the Internet and apply the clustering algorithm. We use the standard color, texture, and edge features of images. The white cells are empty and the black cells correspond to the ants with high fitness values. The dark gray and light gray cells represent lower fitness values in that order. In Table 9.3, the "Location" column has locations of those ants which correspond to darker shades in Fig. 9.7 at the top (after sorting). In our experiments we generally add only the top 20 locations to the index. In this case, $\Omega_1 = \{\varpi_{42}, \varpi_1, \varpi_{28}, \varpi_{300}, \ldots\}$ is an example.

The computational complexity of clustering the ants is $\mathcal{O}(s_x \times s_y \times m \times T)$ where $s_x \times s_y$ is the area neighborhood of each ant, m is the total number of ants, and T is the number of iterations. The computational complexity of the index generation is $\mathcal{O}(m^2)$. Both these computations are done only once and are offline preprocessing steps, i.e., not during query processing. The clustering could generate two or more clusters of ants which which are spatially far apart on the grid even though they are quite similar to one another. The index generation method ensures that similar ants among the different clusters are considered as in the same group. We explain this further in the following subsection.

9.3.2.2 Retrieval Through Relevance Feedback

Now we describe how the generated index using the Algorithm 1 is efficiently used to enable retrieval of matching images using relevance feedback from user.

Algorithm 2 is used to determine the result set of similar images, Δ. As the first step, the search ant ϖ_s is initialized randomly at η locations, and its fitness values at those locations are calculated as shown in Fig. 9.1. For our method to be efficient, $\eta \ll n$. We have verified through our experiments that η required is small enough for our method to show improvements in the overall performance. The ants whose fitness is greater than γ are added to Δ (line 10 of Algorithm 2).

Using Figs. 9.1, 9.2, 9.3, 9.4, and 9.5, we present an arbitrary example to explain the use of the generated index for relevance feedback. Figure 9.1 represents a part of a grid after clustering. Figure 9.3 shows the set of images retrieved when lines 1–12 of the Algorithm 2 are executed. At this stage of the execution,

$$\Delta = \{\varpi_{94}, \varpi_{21}, \varpi_{70}, \varpi_{73}, \varpi_{9}, \varpi_{72}, \varpi_{44}, \varpi_{89}, \varpi_{201}, \varpi_{35}\}$$

For each of these ants, the corresponding similar ants from the index (Ω_l) are shown in the adjacent box. After the first relevance feedback from the user, the Δ is updated to $\{\varpi_{70}, \varpi_{21}\}$ (line 14). Figure 9.1 shows all the images similar to ϖ_s in gray cells. The index is able to connect similar images across multiple clusters, in this case across the *three* clusters.

Now the result Δ is updated with the top κ ants corresponding to the index values of $\{\varpi_{70}, \varpi_{21}\}$, i.e.,

$$\Delta = \{\varpi_{70}, \varpi_{21}, \varpi_{216}, \varpi_{73}, \varpi_{9}, \varpi_{72}, \varpi_{44}, \varpi_{89}, \varpi_{201}, \varpi_{36}\}$$

Figure 9.4 shows the second relevance feedback from the user. The number of matching images has increased to $four$. It can be observed that the relevant ants repeat across the index. These are shown in dark gray boxes. The light gray boxes contain those ants which are the correct matches and should be selected by the user in the next relevance feedback. Observe how the index spans across the three clusters in Fig. 9.1. After the second relevance feedback $\Delta = \{\varpi_{70}, \varpi_{21}, \varpi_{216}, \varpi_{36}\}$. The third relevance feedback in Fig. 9.5 shows that the algorithm was successful (with

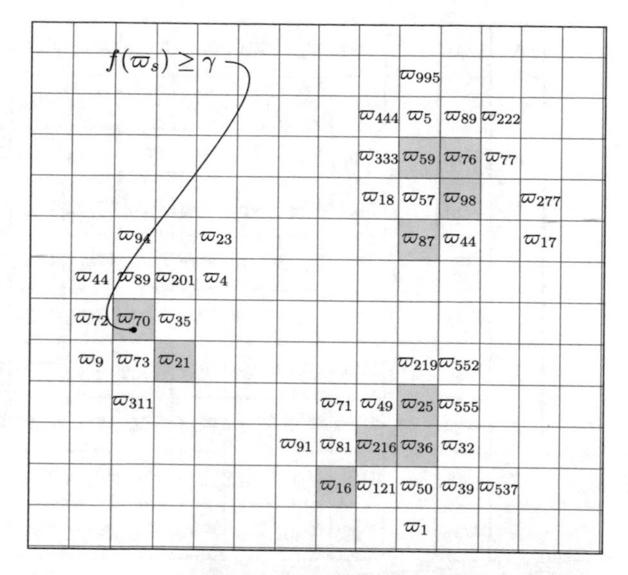

Fig. 9.1 Clustering and retrieval

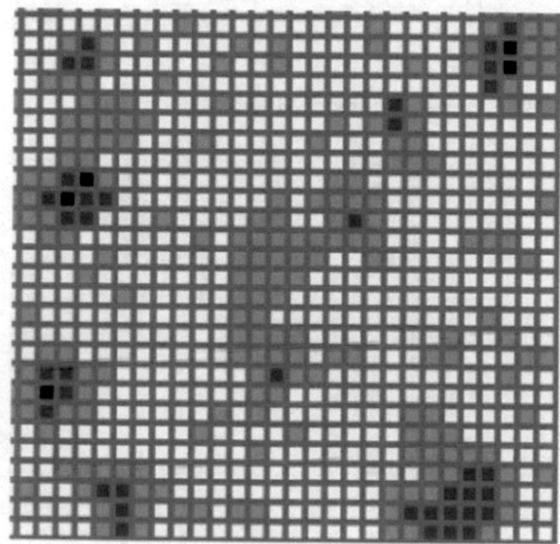

Fig. 9.2 A portion of the grid after clustering 20,000 images

$N_{RF} = 3$) in retrieving the top $\kappa = 10$ matching images. The corresponding index values show a high repetition of the relevant ants in the dark gray boxes.

9.3.3 Extending General Topological Model to ASM-Based RF

In this subsection we extend the mathematical constructs just developed and relate them with the algorithms explained in Sect. 9.3.

Consider a *clustered space*, $\mathfrak{C}$ which contains n ants. The *grid* corresponding to $\mathfrak{C}$ consists of n locations each containing one ant. As the clustering is complete, we consider the grid as frozen, or in other words, an ant does not move to new

Relevance Feedback	
✗	ϖ_{94}
✓	ϖ_{21}
✓	ϖ_{70}
✗	ϖ_{73}
✗	ϖ_{9}
✗	ϖ_{72}
✗	ϖ_{44}
✗	ϖ_{89}
✗	ϖ_{201}
✗	ϖ_{35}

Ω_l

ϖ_{43}	ϖ_{23}	ϖ_{3}	ϖ_{30}	ϖ_{723}	ϖ_{10}	ϖ_{403}	ϖ_{113}	ϖ_{304}	ϖ_{173}	ϖ_{93}
ϖ_{70}	ϖ_{216}	ϖ_{94}	ϖ_{73}	ϖ_{9}	ϖ_{712}	ϖ_{44}	ϖ_{89}	ϖ_{201}	ϖ_{35}	ϖ_{233}
ϖ_{94}	ϖ_{26}	ϖ_{73}	ϖ_{9}	ϖ_{72}	ϖ_{414}	ϖ_{89}	ϖ_{201}	ϖ_{26}	ϖ_{36}	ϖ_{25}
ϖ_{43}	ϖ_{23}	ϖ_{3}	ϖ_{59}	ϖ_{113}	ϖ_{759}	ϖ_{509}	ϖ_{659}	ϖ_{239}	ϖ_{559}	ϖ_{199}
ϖ_{6}	ϖ_{201}	ϖ_{101}	ϖ_{77}	ϖ_{211}	ϖ_{911}	ϖ_{119}	ϖ_{772}	ϖ_{111}	ϖ_{329}	ϖ_{92}
ϖ_{307}	ϖ_{309}	ϖ_{703}	ϖ_{22}	ϖ_{27}	ϖ_{370}	ϖ_{175}	ϖ_{1}	ϖ_{70}	ϖ_{17}	ϖ_{327}
ϖ_{59}	ϖ_{76}	ϖ_{87}	ϖ_{57}	ϖ_{712}	ϖ_{2}	ϖ_{72}	ϖ_{70}	ϖ_{1}	ϖ_{112}	ϖ_{12}
ϖ_{204}	ϖ_{24}	ϖ_{124}	ϖ_{14}	ϖ_{4}	ϖ_{218}	ϖ_{400}	ϖ_{294}	ϖ_{412}	ϖ_{41}	ϖ_{22}
ϖ_{181}	ϖ_{1}	ϖ_{112}	ϖ_{12}	ϖ_{10}	ϖ_{115}	ϖ_{811}	ϖ_{171}	ϖ_{85}	ϖ_{11}	ϖ_{101}
ϖ_{253}	ϖ_{30}	ϖ_{6}	ϖ_{293}	ϖ_{3}	ϖ_{153}	ϖ_{890}	ϖ_{23}	ϖ_{273}	ϖ_{237}	ϖ_{13}

Fig. 9.3 First relevance feedback

Relevance Feedback	
✓	ϖ_{216}
✓	ϖ_{21}
✓	ϖ_{70}
✗	ϖ_{73}
✗	ϖ_{9}
✗	ϖ_{72}
✗	ϖ_{44}
✗	ϖ_{89}
✗	ϖ_{201}
✓	ϖ_{36}

Ω_l

ϖ_{71}	ϖ_{81}	ϖ_{16}	ϖ_{121}	ϖ_{57}	ϖ_{50}	ϖ_{39}	ϖ_{32}	ϖ_{36}	ϖ_{44}	ϖ_{25}
ϖ_{70}	ϖ_{216}	ϖ_{94}	ϖ_{73}	ϖ_{9}	ϖ_{712}	ϖ_{44}	ϖ_{89}	ϖ_{201}	ϖ_{35}	ϖ_{233}
ϖ_{94}	ϖ_{4}	ϖ_{73}	ϖ_{9}	ϖ_{72}	ϖ_{414}	ϖ_{89}	ϖ_{201}	ϖ_{26}	ϖ_{36}	ϖ_{25}
ϖ_{43}	ϖ_{23}	ϖ_{3}	ϖ_{59}	ϖ_{113}	ϖ_{759}	ϖ_{509}	ϖ_{659}	ϖ_{239}	ϖ_{559}	ϖ_{199}
ϖ_{6}	ϖ_{201}	ϖ_{101}	ϖ_{77}	ϖ_{211}	ϖ_{911}	ϖ_{119}	ϖ_{772}	ϖ_{111}	ϖ_{329}	ϖ_{92}
ϖ_{307}	ϖ_{309}	ϖ_{703}	ϖ_{22}	ϖ_{27}	ϖ_{370}	ϖ_{175}	ϖ_{1}	ϖ_{70}	ϖ_{17}	ϖ_{327}
ϖ_{59}	ϖ_{76}	ϖ_{87}	ϖ_{57}	ϖ_{712}	ϖ_{2}	ϖ_{72}	ϖ_{70}	ϖ_{1}	ϖ_{112}	ϖ_{12}
ϖ_{204}	ϖ_{24}	ϖ_{124}	ϖ_{14}	ϖ_{4}	ϖ_{218}	ϖ_{400}	ϖ_{294}	ϖ_{412}	ϖ_{41}	ϖ_{22}
ϖ_{181}	ϖ_{1}	ϖ_{112}	ϖ_{12}	ϖ_{10}	ϖ_{115}	ϖ_{811}	ϖ_{171}	ϖ_{85}	ϖ_{11}	ϖ_{101}
ϖ_{216}	ϖ_{59}	ϖ_{76}	ϖ_{98}	ϖ_{44}	ϖ_{87}	ϖ_{18}	ϖ_{21}	ϖ_{91}	ϖ_{39}	ϖ_{50}

Fig. 9.4 Second relevance feedback

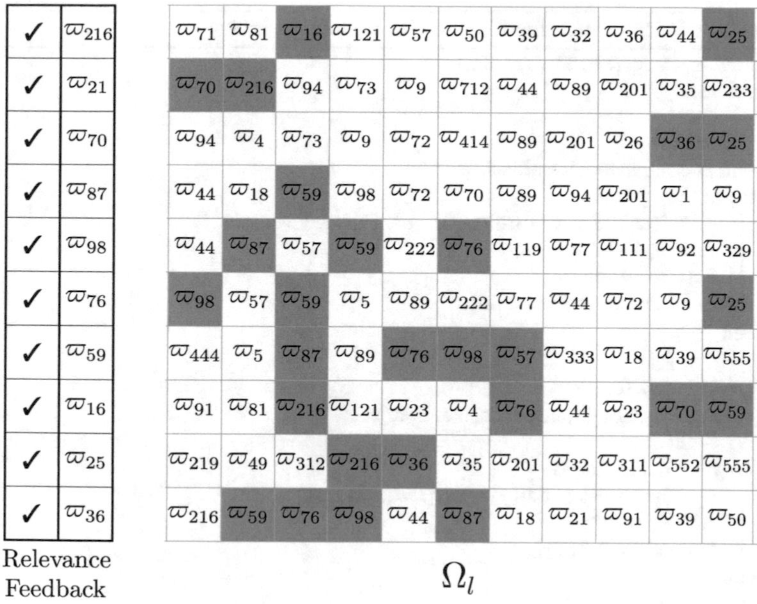

Relevance Feedback	
✓	ϖ_{216}
✓	ϖ_{21}
✓	ϖ_{70}
✓	ϖ_{87}
✓	ϖ_{98}
✓	ϖ_{76}
✓	ϖ_{59}
✓	ϖ_{16}
✓	ϖ_{25}
✓	ϖ_{36}

Ω_l

ϖ_{71}	ϖ_{81}	ϖ_{16}	ϖ_{121}	ϖ_{57}	ϖ_{50}	ϖ_{39}	ϖ_{32}	ϖ_{36}	ϖ_{44}	ϖ_{25}
ϖ_{70}	ϖ_{216}	ϖ_{94}	ϖ_{73}	ϖ_{9}	ϖ_{712}	ϖ_{44}	ϖ_{89}	ϖ_{201}	ϖ_{35}	ϖ_{233}
ϖ_{94}	ϖ_{4}	ϖ_{73}	ϖ_{9}	ϖ_{72}	ϖ_{414}	ϖ_{89}	ϖ_{201}	ϖ_{26}	ϖ_{36}	ϖ_{25}
ϖ_{44}	ϖ_{18}	ϖ_{59}	ϖ_{98}	ϖ_{72}	ϖ_{70}	ϖ_{89}	ϖ_{94}	ϖ_{201}	ϖ_{1}	ϖ_{9}
ϖ_{44}	ϖ_{87}	ϖ_{57}	ϖ_{59}	ϖ_{222}	ϖ_{76}	ϖ_{119}	ϖ_{77}	ϖ_{111}	ϖ_{92}	ϖ_{329}
ϖ_{98}	ϖ_{57}	ϖ_{59}	ϖ_{5}	ϖ_{89}	ϖ_{222}	ϖ_{77}	ϖ_{44}	ϖ_{72}	ϖ_{9}	ϖ_{25}
ϖ_{444}	ϖ_{5}	ϖ_{87}	ϖ_{89}	ϖ_{76}	ϖ_{98}	ϖ_{57}	ϖ_{333}	ϖ_{18}	ϖ_{39}	ϖ_{555}
ϖ_{91}	ϖ_{81}	ϖ_{216}	ϖ_{121}	ϖ_{23}	ϖ_{4}	ϖ_{76}	ϖ_{44}	ϖ_{23}	ϖ_{70}	ϖ_{59}
ϖ_{219}	ϖ_{49}	ϖ_{312}	ϖ_{216}	ϖ_{36}	ϖ_{35}	ϖ_{201}	ϖ_{32}	ϖ_{311}	ϖ_{552}	ϖ_{555}
ϖ_{216}	ϖ_{59}	ϖ_{76}	ϖ_{98}	ϖ_{44}	ϖ_{87}	ϖ_{18}	ϖ_{21}	ϖ_{91}	ϖ_{39}	ϖ_{50}

Fig. 9.5 Third relevance feedback

locations. Let $\mathfrak{A}$ be the set of all ants and $\mathfrak{L}$ be the set of all locations where one ant exists in the frozen *grid*. Consider the product set $\mathfrak{A} \times \mathfrak{L}$, which contains n^2 elements.

Using the notation given in Sect. 9.3, let us define the fitness value $f^*(\varpi_i, \delta_m)$ as $f(\varpi_i)$@ Location δ_m where $(\varpi_i, \delta_m) \in \mathfrak{A} \times \mathfrak{L}$. Fitness is calculated using the Eq. (9.1) by replacing $N(\varpi_i)$ with $N(\varpi_m)$.

$$f^*(\varpi_i, \delta_m) = max \left\{ 0, \frac{1}{S} \sum_{\varpi_j} \left(1 - \frac{d(\varpi_i, \varpi_j)}{\alpha} \right) \right\} \tag{9.4}$$

where

$$\varpi_j \in N(\varpi_m)$$

It is the same as the last column of Table 9.3 having n^2 rows. Let us define a threshold γ such that all the fitness values above γ indicate a similarity between the ant and the neighboring ants at the location where the fitness is calculated. In other words, if $f^*(\varpi_i, \delta_m) > \gamma$, then ϖ_i is similar to any ant $\varpi \in N(\varpi_m)$ including ϖ_m.

Every ant ϖ has n^2 fitness values for the locations. For every ant ϖ_i, we define a set of ants say $A^{\gamma}_{\varpi_i}$ such that for any ant $\varpi_k \in A^{\gamma}_{\varpi_k}$, $f^*(\varpi_i, \delta_k) > \gamma$ where $k \in [1, n]$. This is the generation of index from Algorithm 1. We

Algorithm 2: Relevance feedback using generated index

Input: $\mathfrak{C}$, ϖ_s, γ, κ, $\{\Omega_1, \Omega_2, \ldots, \Omega_n\}$.
Output: Δ
$\Delta \leftarrow \emptyset$;
for $i \leftarrow 1 : \eta$ **do**
  Place ϖ_s at random location δ_r in $\mathfrak{C}$;
  if $f(\varpi_s) \geq \gamma$ **then**
    // $f(\varpi_s)$ is fitness of ϖ_s at location δ_r.
    Add ϖ_r and $N(\varpi_r)$ to Δ;
    Sort Δ over fitness;
    $\Delta \leftarrow$ first κ elements of Δ;
  end
end
for $m \leftarrow 1 : N_{RF}$ **do**
  Update Δ using relevance feedback from user;
  $\Delta^* \leftarrow \emptyset$;
  for $j \leftarrow 1 : size\{\Delta\}$ **do**
    $l \leftarrow$ the position of $\Delta(j) \in \{\varpi_1, \varpi_2, \ldots, \varpi_n\}$;
    Add Ω_l to Δ^*;
    Sort Δ^* over fitness.;
  end
  $\Delta \leftarrow$ first κ elements of Δ^*;
end

construct a topology $\mathbb{T}_{ASM}$ using the arbitrary union and intersection of the n sets $A^{\gamma}_{\varpi_1}, A^{\gamma}_{\varpi_2}, A^{\gamma}_{\varpi_3} \ldots A^{\gamma}_{\varpi_n}$.

Using the notation of Sect. 9.2, we represent our method as $\mathscr{A}_{ASM}$. The relevance feedback mechanism is represented below:

$$\xrightarrow[initial\ query]{i_q} \mathscr{A}^0_{ASM} \xrightarrow[initial\ result]{I^*_0} \text{User} \xrightarrow[rel.fb]{I_{r1}} \mathscr{A}^1_{ASM} \xrightarrow{I^*_1} \text{User} \xrightarrow[rel.fb]{I_{r2}} \ldots \mathscr{A}^n_{ASM} \xrightarrow{I^*_n} \text{User}.$$

The initial result I^*_0 is the result Δ before any of the updates from the Algorithm 2 of the Sect. 9.3. The ants basically represent the images, so for all k, $I^*_k \in \mathbb{T}_{ASM}$.

We define *precision* and *recall* as in [49] below:

$$precision = \frac{|\ \{relevant\ images\} \cap \{retrieved\ images\}\ |}{|\ \{retrieved\ images\}\ |} \tag{9.5}$$

$$recall = \frac{|\ \{relevant\ images\} \cap \{retrieved\ images\}\ |}{|\ \{relevant\ images\}\ |} \tag{9.6}$$

Lemma 1 *The precision of I^*_n for a query image i_q is 1 only if $\{A^{\gamma}_{\varpi_1} \cup A^{\gamma}_{\varpi_2} \cup A^{\gamma}_{\varpi_3} \ldots \cup A^{\gamma}_{\varpi_n}\} \subseteq \mathscr{D}$.*

$f_{\text{orc}}(S_q) = \mathscr{D}$ *(from Sect. 9.2).*

Proof From the Eq. 9.5, $\mathscr{D} = \{relevant\ images\}$. For *recall* to be equal to **1**, the numerator should be equal to the denominator which is possible only if $\{retrieved\ images\}$ is contained within the union $\{A^{\gamma}_{\varpi_1} \cup A^{\gamma}_{\varpi_2} \cup A^{\gamma}_{\varpi_3} \ldots \cup A^{\gamma}_{\varpi_n}\}$.

Lemma 2 *The recall of I^*_n for a query image i_q is **1** only if $\{A^{\gamma}_{\varpi_1}, A^{\gamma}_{\varpi_2}, A^{\gamma}_{\varpi_3} \ldots A^{\gamma}_{\varpi_n}\}$ is a **cover** of $\mathscr{D}$.*

Proof By definition, a **cover** of a set A in a topological space is a family of sets $\{U_\alpha\}$ such that $A \subseteq \cup_\alpha U_\alpha$. I^*_x, the output from $\mathscr{A}^x_{ASM}$, is always a subset of the arbitrary union of the sets $\{A^{\gamma}_{\varpi_1}, A^{\gamma}_{\varpi_2}, A^{\gamma}_{\varpi_3} \ldots A^{\gamma}_{\varpi_n}\}$. The retrieval is accurate if $I^*_n = \mathscr{D}$ and $I^*_n \subseteq \cup_\alpha A^{\gamma}_\alpha$; so $\mathscr{D} \subseteq \cup_\alpha A^{\gamma}_\alpha$ if $recall = 1$.

Now we construct a theorem to compare two algorithms $\mathscr{A}_K$ and $\mathscr{A}_L$ for relevance feedback. We review some of the definitions which we will use in the theorem. A set $\mathscr{N}^x$ is called a **neighborhood** of a point x of the topological space $(\mathscr{I}, \mathscr{T}_{\mathscr{I}})$ if there exists an open set $V \in \mathscr{I}$ such that $x \in V$ and $V \in \mathscr{N}^x$. A point x is called a **contact point** of a set $A \subset (\mathscr{I}, \mathscr{T}_{\mathscr{I}})$ if every *neighborhood* of x contains at least one point of A. The set of all *contact points* of a set $A \subset (\mathscr{I}, \mathscr{T}_{\mathscr{I}})$ is called the **closure** of A, denoted by $\overline{A}$. The term "decision boundary" used in the following theorem is the same as used in general classification problems.

Theorem 1 *For any two algorithms used for the relevance feedback, say $\mathscr{A}_L$ and $\mathscr{A}_K$, in the m_{th} relevance feedback, the algorithm $\mathscr{A}_L$ is "same or better" than $\mathscr{A}_K$ if $|\ \overline{I^*_{Lm}}\ | > |\ \overline{I^*_{Km}}\ |$, when the* recall *(Eq. 9.6) is the same. $\mathscr{A}_K$ and $\mathscr{A}_L$ are assumed to be based on nearest neighbor search (NNS). "Same or better" is used in the context of relevance feedback.*

Proof Without loss of generality, we can state that for a nearest neighbor search algorithm used for image retrieval, if $E \cap f_{\text{orc}}(S_q) \neq \phi$, where $E \in (\mathscr{I}, \mathscr{T}_{\mathscr{I}})$, if $x \in E$ and $y \in \mathscr{N}^x$, a neighborhood of x, then y is a good candidate for being on the same side of the "decision boundary" as x.

At the mth relevance feedback, for calculating recall, $\{retrieved\ images\} = I^*_{(K,L)m}$ for each algorithm and $\{relevant\ images\} = f_{\text{orc}}(S_q)$ where S_q is the set of user's semantic concepts. So if the recall is same, then $|\ f_{\text{orc}}(S_q) \cap I^*_{Lm}\ | = |\ f_{\text{orc}}(S_q) \cap I^*_{Km}\ |$. From the above reasoning, for the $m+1$th relevance feedback, $I^*_{L(m+1)}$ will have more good candidates than $I^*_{K(m+1)}$ if $|\ \overline{I^*_{Lm}}\ | > |\ \overline{I^*_{Km}}\ |$. This follows from the definition of a *contact point*. So if the neighborhoods of a contact point, say x, contain elements of E, then those neighborhoods, which as per our model all belong to $(\mathscr{I}, \mathscr{T}_{\mathscr{I}})$, provide more good candidates for relevance feedback from the user, because the number of contact points in the result from the algorithm $\mathscr{A}_L$ is higher.

In a large database, the time required to seek an image from the storage device is important. We employ the better characteristics of Algorithm $\mathscr{A}_L$ to efficiently store the images on storage device. Using the fact that files stored closer to each other in the same segment of a storage device are faster to seek, we formally state the following corollary.

Corollary 1 *For the two algorithms mentioned in Theorem* 1, *the average time of response (from storage device) for* $\mathscr{A}_L$ *will be same or lower than* $\mathscr{A}_K$, *during relevance feedbacks.*

Proof Since $| \overline{I^*_{Lm}} | > | \overline{I^*_{Km}} |$, if these sets are stored in the same segment of a storage device, the algorithm $\mathscr{A}_L$ will show lower average time of response. $\mathscr{A}_L$ is better as it enables storing of larger number of "similar" images within the same segment. In the worst case, if none of the images are relevant, the seek times will be similar for both the algorithms.

We use $| \overline{I^*_{Lm}} |$ from the ASM technique, to generate the index for image datasets and compare it with other methods for response times from storage device in Sect. 9.4.

9.4 Experiments

In this section, we first describe how to experimentally determine the optimal parameters for clustering. Figures 9.6 and 9.7 show the clustered spaces before and after clustering, respectively. We used 5000 images randomly chosen from www.flickr.com for clustering and employed color- and edge-based histograms as descriptors for generating input vectors. Each input vector (ant) is shown as a square. Initially, Fig. 9.6 shows the unclustered space where ants are randomly placed. We used $S_x = 1$, $S_y = 1$, $\beta = 0.1$, $\lambda = 2.0$, and total iterations = 20,000 as constants for clustering.

We define *mean fitness* as the average of fitness values $(f(\varpi_i))$ of ants over the entire population, i.e., *mean fitness* = $\frac{\sum_i f(\varpi_i)}{n}$. Mean fitness is the indicator of the progress of clustering. As the number of completed iterations increases, the value of mean fitness also increases, with more and more similar ants being

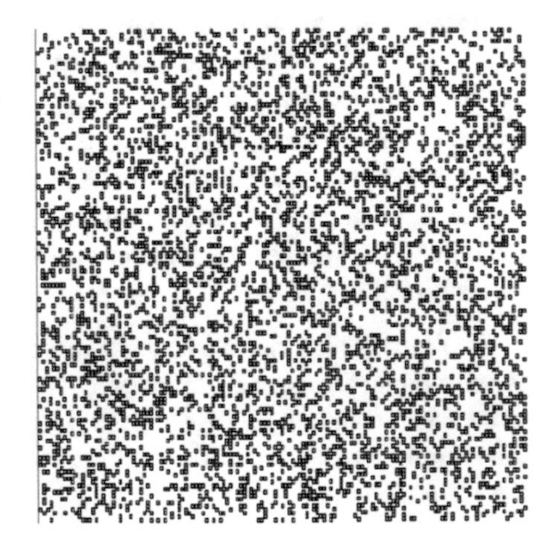

Fig. 9.6 Unclustered space

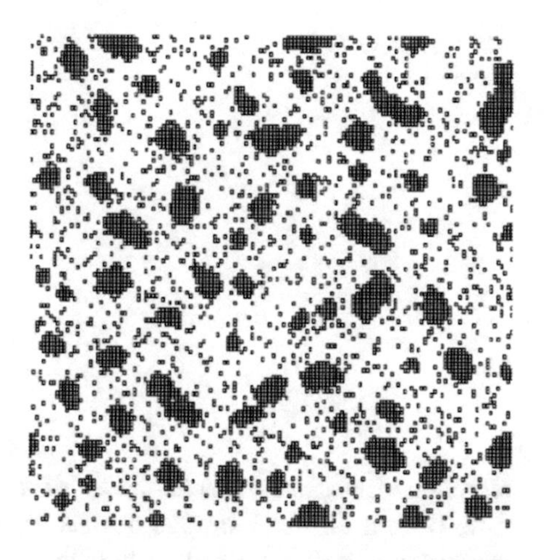

Fig. 9.7 Clustered space

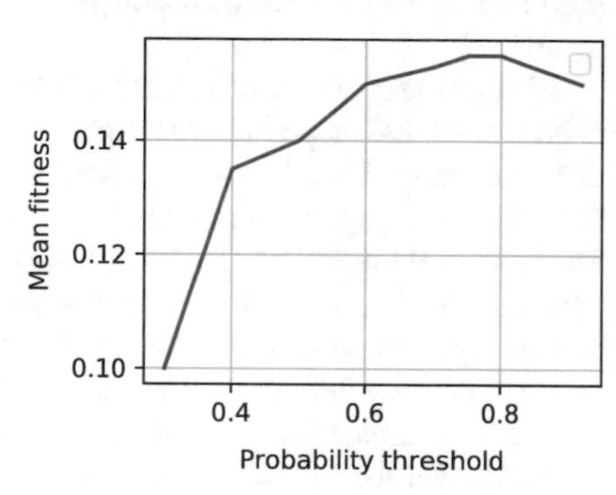

Fig. 9.8 Mean fit. vs prob. threshold

grouped together. As shown in Fig. 9.8 (see the definition of active probability p in Sect. 9.3), a probability threshold of 0.8 is found to be suitable for achieving high mean fitness. We have observed this value to hold good in many experiments. Similarly, the number of iterations vs. mean fitness for 20,000 random images are shown in Fig. 9.9. We generally use a high number of iterations to ensure that we get maximum possible mean fitness values, though we have visually observed no changes in the plot of clustered space (e.g., Fig. 9.7) once the mean fitness stabilizes (e.g., 1500 iterations in Fig. 9.9). For clustering of 50,000 images, we usually run the A^4C algorithm for 20,000 iterations.

9.4.1 *Performance Metric*

When we try to solve the problem of building an efficient index to scale up an image retrieval system [70], it is also important to determine how the index structures work with the storage systems. This is more significant when the size of the image database is very large and retrieval is done across millions of images. Our proposed

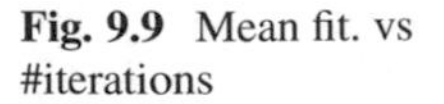

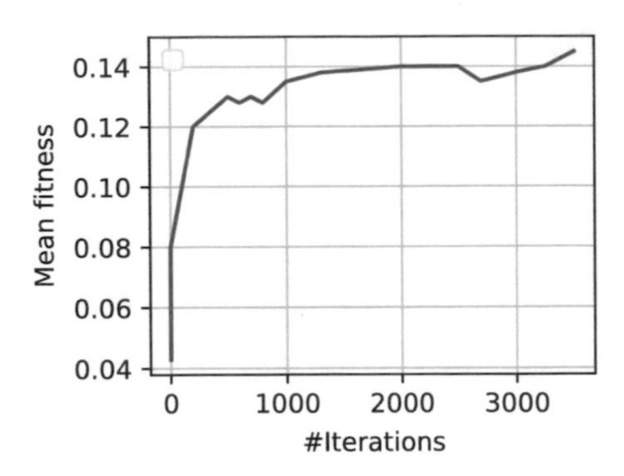

Fig. 9.9 Mean fit. vs #iterations

technique intuitively enables us to use the generated index to reduce the retrieval time from a physical storage device. In this section, we compare our work using performance metrics based on storage disk access times and which shows promising results.

We use the performance metric of Ramaswamy and Rose [51]. Accessing data from a hard disk is possible through serial access and random access, which have different costs. Data is always stored and retrieved in *pages*. The time spent in accessing one page with random access is significantly greater than serial access. The proposed metric in [51] shows that between two indexing schemes, the one for which the number of random access pages used is less for the same number of serial access pages in both is better than the other (refer to [51] for more details). We verify the theory developed (Corollary 1) before, by comparing the performance of the index generated from our algorithm against the other methods used for relevance feedback over the following data sets.

9.4.2 Dataset: Cortina

This dataset contains 1,103,271 elements from CORTINA [14] and consists 48-D MPEG-7 texture features. We divided the dataset into 50 parts for clustering but produced a single index for all clusters. We use the same constants ($S_x = 1$, $S_y = 1$, $\beta = 0.1$, $\lambda = 2.0$, and total iterations = 20,000) as used previously.

We typically use 20,000 ants in one grid for clustering. With more powerful computational units, the number of ants used in one grid could be increased. When the size of the grid (number of ants used in one instance of clustering) is increased, the similar ants which must ideally belong to the same cluster form smaller clusters instead and are spread out in the grid. The values of mean fitness were observed to reduce when larger number of ants (>20,000) were used concurring with the above statement.

We compared the I/O performance and the computational performance of our technique with [51]. Figure 9.10 shows the I/O performance using their performance metric described above. We refer to our indexing technique as ASM-indexing in

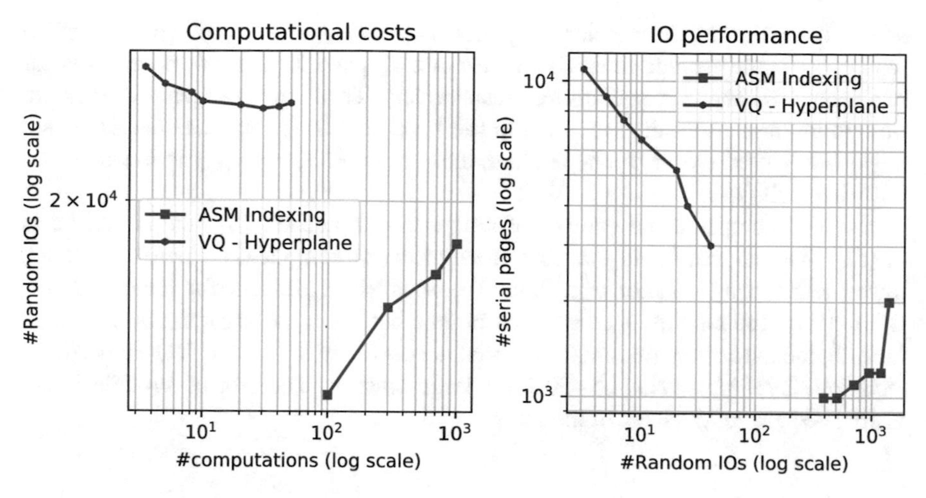

Fig. 9.10 Results for Cortina dataset

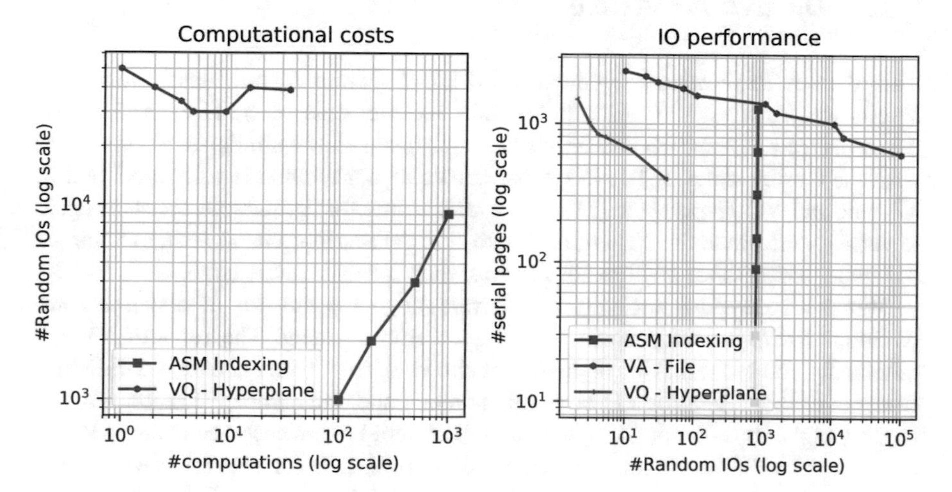

Fig. 9.11 Results for Bio-Retina dataset

Figs. 9.10 and 9.11. The I/O performance is shown with varying cluster sizes. Even though, in ASM-indexing, the number of random I/Os seems slightly higher, it must be noted that the random access is actually confined to a single cluster which is, in practice, placed closer in a hard disk. Therefore, the seek time is actually limited. With this noted, the number of serial access used is mostly constant due to the nature of our indexing (refer to Algorithm 1). Thus for large databases, our indexing technique is advantageous.

When we say that random access is confined within the same cluster, we refer to Algorithm 2. The initial Δ obtained with the η placements is over only one grid (20,000 ants in this experiment), and the index is used to find similar ants (images) across the 50 grids. Furthermore, suppose $T_{\text{res}} = T_{\text{seq}}S + T_{\text{rand}}R$ (refer to [51]) where T_{res} is the average time of response and T_{seq} and T_{rand} are average times for one sequential and one random I/O, respectively. Then since the random I/Os are limited to one cluster, the seek time in the hard disk is less because the pages are physically stored closer to one another, and so the effective T_{rand} is lower in our indexing technique.

We also compared the number of computations needed with the experiments in [51]. We find that for this dataset, the number of computations needed is much lower, reduced by a factor of at least 10, which is significant for large datasets. Once again, the random access seems higher but the same argument holds (seek times in the hard disks are low). The serial access is lower than the VQ-Hyperplane technique by a high value of 450% for large clusters. The plot of VA-File is not visible because of the scale of the plots.

9.4.3 Dataset: Bio-Retina

This dataset Bio-Retina [5] contains MPEG-7 texture feature descriptors. The results are shown in Fig. 9.11. It has 208,506 elements of 62-D each. The I/O performance based on the chosen metric once again shows similar performance as with the Cortina dataset. The I/O performance of our method is compared against VA-File and VQ-Hyperplane [51]. The average time for the random access is again lower for ASM-indexing. Therefore, with this dataset too, our indexing technique has good performance and has advantages when the database is large.

When we compare our results with respect to the number of distance computations, ASM-indexing again shows good improvement. The random IOs are practically constant for ASM-indexing. It shows an improvement over VA-File by a factor of 100. Though for VQ-Hyperplane technique, the number of random IOs is slightly higher by a factor of 10, the number of serial pages accessed is much lower. It is lower by 400% for smaller clusters and up to 15% for clusters of maximum size. The computational cost is also lower by at least 5 $times$ for large clusters when compared to VQ-Hyperplane technique.

9.4.4 Comparison with PicSOM

9.4.4.1 ASM and PicSOM

PicSOM [28, 34] is a CBIR system which utilizes tree-structured self-organizing maps. It applies relevance feedback toward "the weighing approach" (as explained in the Online Appendix) to improve query results. Self-organizing maps are intrinsically similar to the ant clustering techniques [25–27]. We provide a theoretical comparison of the two techniques and show their similarity. We also quantitatively compare the proposed system against an online implementation of PicSOM and prove the efficacy of our system for relevance feedback.

The notation used in this section is consistent with [26] and is used only in this subsection.

Ant-based clustering (including clustering based on ASM) operates on a low-dimensional grid $\mathscr{G} \subset \mathbb{N}^2$, where $\mathbb{N}$ is the set of natural numbers. A finite set of input vectors $\mathscr{X}$ is projected onto $\mathscr{G}$ by $m : \mathscr{X} \rightarrow \mathscr{G}$. The mapping is altered by the ants as they move to neighboring grid nodes, thus moving the input vector $x \in X$ from $m(x)$ to $m'(x)$. This movement or distortion is represented by ϕ in (9.7) where the neighborhood to which the ant can move is $N_x(i) = \{y \in X : y \neq x, m(y)$ neighboring $i\}$.

$$\phi_x(i) = \frac{1}{|N_x(i)|} \sum_{y} \left(1 - \frac{||x - y||}{\alpha}\right) \tag{9.7}$$

where

$$y \in N_x(i)$$

Similarly, self-organizing maps consist of a grid $\mathscr{G}$ over which the mapping function $m : \mathscr{X} \rightarrow \mathscr{G}$ iteratively updates the position of input vectors $\mathscr{X}$. The error function for SOM is shown in (9.8) where $h : \mathscr{G} \times \mathscr{G} \rightarrow [0, 1]$ is used to produce the codebook vectors (for more details, refer to [31]).

$$\Phi_x(i) = \frac{|| \sum_{y \in X} h(m(y), i) \cdot (x - y)||}{\sum_{y \in X} h(m(y), i)} \tag{9.8}$$

It has been explained in [26] that $|N_x(i)| = \sum_{y \in X} h(m(y), i)$, which leads to (9.9). Thus the ant clustering technique is (as stated in [26]) a non-deterministic derivative of the SOM.

$$\phi_x(i) = \frac{1}{|N_x(i)|} \sum_{y \in N_x(i)} \left(1 - \frac{\Phi_x(i)}{\alpha}\right) \tag{9.9}$$

PicSOM uses SOM as the indexing method, with the SOM layers forming a hierarchical tree structure (for computational benefits). The relevance feedback from the user is applied to the SOM layers on which similar images are arranged near each other. In the ASM model, we use the same fact that the similar images are arranged together over the grid, but unlike PicSOM, which identifies the best matching units [34] and involves computation during relevance feedback, ASM uses the error function (9.7) for indexing and involves no distance computation at the time of search. The index ensures that related clusters, even if separated spatially on the grid ($\mathscr{G}$), do not affect retrieval.

9.4.4.2 Experimental Comparison with PicSOM

We compare the ASM retrieval with PicSOM [32] using their online implementation at [57]. We scraped 4348 images from [57] and extracted the three different features as described in [7, 32]. We briefly explain the image features used. Color, texture, and shape features were extracted from the images. Color feature is a 15-dimensional vector which is the average of RGB values calculated in five zones [33] of the images, corresponding to the left, right, top, bottom, and center parts of the image area. Texture features are calculated likewise in the five zones. For texture features, each pixel's YIQ luminance is compared to that of its eight neighbors. The estimated probability that the center pixel has a larger value than the neighbor is used as a feature. It is a 40-dimensional vector. For extracting shape features [7], we transform the images to binarized edge images using 3×3-sized Sobel masks. The edge directions are then discretized to eight values, and an 8-bin histogram is formed from the directions in the same five zones. Shape feature is a 40-dimensional vector.

We compare the retrieval performance of ASM over the dataset using the same method as in [32]. Aircrafts, buildings, and human faces are the three classes of images which are used in the experiments. For example, for each of the 348 aircraft images, the total number of images shown by ASM until that particular image is shown is counted programmatically. The average number of images shown before the hit is divided by total number of images (4348) which yields a value $\tau \in (0, 1]$. $\tau = 0.5$ is the baseline for random picking. In our experiments, 16 new images are presented at every iteration of retrieval, same as in [32].

We describe our experiment in more detail in conjunction with the theory presented in Sect. 9.3.3. Since we experiment on three features and a combination of all of them, we develop four *clustered spaces*. Now consider a set of 348 aircraft images; each of the 348 ants has a corresponding set $A_{\varpi_i}^{T_f}$. We used the threshold $T_f = 0$ for all the features. To generate $A_{\varpi_i}^{T_f}$, we use Algorithm 1. While generating $A_{\varpi_i}^{T_f}$, we also add all the neighbors of the ant from the *clustered space*, even if for some cases $T_f = 0$. For a target ant (ϖ_t) to get hit during iterations of retrieval, it is imperative that it be a member of at least one or more $A_{\varpi_j}^{T_f}$ where $\varpi_j \in A_{\varpi_t}^{T_f}$, i.e., the index corresponding to the ant ϖ_t. There are some instances when this is not the case and the target image is never retrieved. When such an instance occurs, we calculate τ by adding half of the remaining images in the dataset to the count. For example, if 1205 images are already shown and the target image is not retrieved, then we use $1205 + (4348 - 1205)/2$ as the final count to calculate tau which is equivalent to picking the target image randomly from the remaining images. Such instances are observed for most of the images when $texture$ feature is used. On the other hand no such instance occurs while the $shape$ feature is used.

We make the following conclusions from the results of our experiments shown in Table 9.4.

Table 9.4 Retrieval performances τ for ASM and PicSOM using three distinct features and all the three combined. Three different image classes are used

		Color	Texture	Shape	All
	Aircraft	0.307	0.436	0.220	0.261
PicSOM	Buildings	0.319	0.312	0.309	0.300
	Faces	0.372	0.356	0.364	0.362
	Mean	0.333	**0.368**	0.298	0.308
	Aircraft	0.263	0.580	0.080	0.265
ASM	Buildings	0.270	0.421	0.104	0.272
	Faces	0.313	0.458	0.095	0.286
	Mean	**0.282**	0.486	**0.093**	**0.274**

- ASM performs better than PicSOM. On an average ASM shows 13% improvement over PicSOM for this dataset. In the case of shape features, the improvement shown is as high as 68%.
- Texture feature performs poorly, and shape feature gives the best results for both ASM and PicSOM. PicSOM on the one hand shows exceptional improvement in retrieval using shape feature, but the poor results from texture feature are exacerbated on the other hand.
- Both ASM and PicSOM perform poorly in retrieval for the image class of human faces as compared to the other two classes.

9.5 Future Directions and Conclusion

In summary, we have developed a mathematical model to analyze semantic image retrieval systems and used it to develop a nature-inspired image retrieval algorithm based on relevance feedback. We extended our mathematical model to analyze the proposed algorithm theoretically. We were able to define the performance metrics used for image retrieval using our proposed mathematical model. In general, nature-inspired algorithms face a criticism that they lack theoretical underpinnings, and due to this reason, they are seldom used for practical applications. To this end, our general topology-based model of image retrieval using relevance feedback is an important step. Furthermore, our proposed ant-based image retrieval algorithm has been compared with existing approaches and shown to be more efficient. In particular, our approach compared well against PicSOM both theoretically and empirically.

In real world, apart from visual cues, textual information, click logs, speech information in videos, etc. are simultaneously available and should be used collaboratively to retrieve useful information. For example, if textual and visual cues are to be considered together, then it requires a collaborative indexing approach. Furthermore, for such environments with complex characteristics, nature-inspired algorithms present with easier and more effective solutions. Therefore, we would extend our mathematical model to such cross-modal retrieval systems [63] and further develop our ant-based system toward this goal. Furthermore, quality of query formation is very important for good performance of retrieval systems. We would extend our model to incorporate this aspect of the retrieval system which is not considered very often in the current approaches. Finally, since it is our endeavor to enable nature-inspired algorithms to be used for real-world problems, it is important to compete with current approaches on large number of datasets. In particular, real-world datasets from industry [73], viz., Microsoft Image Grand Challenge on Image Retrieval [43] and Alibaba Large-scale Image Search Challenge [1], would be considered to demonstrate the effectiveness of our approaches.

References

1. Alibaba large-scale image search challenge (2015) https://tianchi.aliyun.com/competition/information.htm?raceId=231510
2. Aly M, Munich M, Perona P (2011) Indexing in large scale image collections: scaling properties and benchmark. In: 2011 IEEE Workshop on Applications of Computer Vision (WACV). IEEE, pp 418–425
3. Basener W (2006) Topology and its applications. Pure and applied mathematics: a Wiley series of texts, monographs and tracts. Wiley, Hoboken. http://books.google.co.in/books?id=hn8y3dCP2c8C
4. Bellman RE (1961) Adaptive control processes – a guided tour. Princeton University Press, Princeton
5. Bioretina dataset (2013) http://www.scl.ece.ucsb.edu/datasets
6. Bonnlander BV, Weigend AS (1994) Selecting input variables using mutual information and nonparametric density estimation, in Proceedings of the 1994 International Symposium on Artificial Neural Networks (ISANN'94), pp 42–50
7. Brandt S, Laaksonen J, Oja E (1999) Statistical shape features for content-based image retrieval. J Math Image Vision 17(2):187–198
8. Burgin M, Eberbach E (2013) Ubiquity symposium: evolutionary computation and the processes of life: perspectives and reality of evolutionary computation: closing statement. Ubiquity 2013:5:1–5:12. https://doi.org/10.1145/2555235.2555240
9. Califano A, Mohan R (1994) Multidimensional indexing for recognizing visual shapes. IEEE Trans Pattern Anal Mach Intell 16:373–392
10. Castelli V (2001) Multidimensional indexing structures for content-based retrieval. Technical report, IBM
11. Chen JY, Bouman CA, Allebach JP (1997) Fast image database search using tree-structured VQ. Int Conf Image Process 2:827
12. Chen JY, Bouman CA, Member S, Dalton JC (2000) Hierarchical browsing and search of large image databases 9:442–455
13. Chen L, Xiaohua X, Yixin C, Ping H (2004) A novel ant clustering algorithm based on cellular automata. In: IAT'04: Proceedings of the Intelligent Agent Technology, IEEE/WIC/ACM International Conference. IEEE Computer Society, Washington, DC, pp 148–154
14. Cortina dataset (2013) http://cortina.ece.ucsb.edu/
15. Cox IJ, Miller ML, Minka TP, Papathomas TV, Yianilos PN (2001) The bayesian image retrieval system, pichunter: theory, implementation, and psychophysical experiments. IEEE Trans Image Process 9(1):20–37
16. Datta R, Joshi D, Li J, Wang JZ (2008) Image retrieval: ideas, influences, and trends of the new age. ACM Comput Surv 40:5:1–5:60

17. Dorigo M, Gambardella LM (1997) Ant colony system: a cooperative learning approach to the traveling salesman problem. IEEE Trans Evol Comput 1(1):53–66
18. Dorigo M, Maniezzo V, Colorni A (1996) Ant system: optimization by a colony of cooperating agents. IEEE Trans Syst Man Cybern B (Cybernetics) 26(1):29–41
19. Ferhatosmanoglu H, Tuncel E, Agrawal D, El Abbadi A (2000) Vector approximation based indexing for non-uniform high dimensional data sets. In: Proceedings of the Ninth International Conference on Information and Knowledge Management, CIKM'00. ACM, New York, pp 202–209
20. Floridi L (2014) Semantic conceptions of information. In: Zalta EN (ed) The Stanford encyclopedia of philosophy, Spring 2017 Edition. https://plato.stanford.edu/archives/spr2017/entries/information-semantic/
21. Friedman JH (1997) On bias, variance, 0/1 loss, and the curse-of-dimensionality. Data Min Knowl Disc 1(1):55–77
22. Gudivada VN, Raghavan VV (1995) Content based image retrieval systems. Computer 28(9):18–22
23. Guttman A (1984) R-trees: a dynamic index structure for spatial searching. SIGMOD Rec 14:47–57
24. Handl J, Knowles J, Dorigo M (2006) Ant-based clustering and topographic mapping. Artif Life 12(1):35–62
25. Herrmann L, Ultsch A (2008) The architecture of ant-based clustering to improve topographic mapping. In: Proceedings of the 6th International Conference on Ant Colony Optimization and Swarm Intelligence, ANTS'08. Springer, Berlin/Heidelberg, pp 379–386
26. Herrmann L, Ultsch A (2008) Explaining ant-based clustering on the basis of self-organizing maps. In: Proceedings of the European Symposium on Artificial Neural Networks (ESANN 2008), pp 215–220
27. Herrmann L, Ultsch A (2010) Strengths and weaknesses of ant colony clustering. In: Fink A, Lausen B, Seidel W, Ultsch A (eds) Advances in data analysis, data handling and business intelligence, studies in classification, data analysis, and knowledge organization. Springer, Berlin/Heidelberg, pp 147–156
28. Jorma L, Markus K, Sami L, Erkki O (2000) PicSOM–content-based image retrieval with self-organizing maps. Pattern Recogn Lett 21(13–14):1199–1207
29. Kennedy J (2011) Particle swarm optimization. In: Sammut C, Webb GI (eds) Encyclopedia of machine learning. Springer, New York, pp 760–766
30. Kherfi M, Ziou D (2006) Relevance feedback for CBIR: a new approach based on probabilistic feature weighting with positive and negative examples. IEEE Trans Image Process 15(4):1017–1030
31. Kohonen T, Schroeder MR, Huang TS (eds) (2001) Self-organizing maps, 3rd edn. Springer, New York/Secaucus
32. Laaksonen J, Koskela M, Oja E (1999) Picsom: self-organizing maps for content-based image retrieval. In: International Joint Conference on Neural Networks. IJCNN'99, Washington, DC, vol 4. IEEE, pp 2470–2473
33. Laaksonen JT, Koskela JM, Oja E (1999) Picsom – a framework for content-based image database retrieval using self-organizing maps. In: In 11th Scandinavian Conference on Image Analysis, Kangerlusssuaq, pp 151–156
34. Laaksonen J, Member A, Koskela M, Oja E (2002) Picsom self-organizing image retrieval with MPEG-7 content descriptions. In: Guan L (ed) Networks, special issue on intelligent multimedia processing. IEEE, New York, pp 841–853
35. Le Hégarat-Mascle S, Kallel A, Descombes X (2007) Ant colony optimization for image regularization based on a nonstationary Markov modeling. IEEE Trans Image Process 16(3):865–878
36. Liu J, Tsui KC (2006) Toward nature-inspired computing. Commun ACM 49(10):59–64
37. Liu J, Tang YY, Cao Y (1997) An evolutionary autonomous agents approach to image feature extraction. IEEE Trans Evol Comput 1(2):141–158

38. Liu Y, Zhang D, Lu G, Ma WY (2007) A survey of content-based image retrieval with high-level semantics. Pattern Recognit 40(1):262–282
39. Liu D, Yan S, Ji RR, Hua XS, Zhang HJ (2013) Image retrieval with query-adaptive hashing. ACM Trans Multimedia Comput Commun Appl 9(1):2:1–2:16. https://doi.org/10.1145/2422956.2422958
40. MacArthur S, Brodley C, Shyu C (2000) Relevance feedback decision trees in content-based image retrieval. In: Proceedings of the IEEE Workshop on Content-Based Access of Image and Video Libraries (CBAIVL'00). IEEE Computer Society, Washington, DC, pp 68–
41. Martens D, De Backer M, Haesen R, Vanthienen J, Snoeck M, Baesens B (2007) Classification with ant colony optimization. IEEE Trans Evol Comput 11(5):651–665
42. Meilhac C, Nastar C (1999) Relevance feedback and category search in image databases. In: Proceedings of the IEEE International Conference on Multimedia Computing and Systems – vol 2, ICMCS'99. IEEE Computer Society, Washington, DC, pp 512–517
43. Microsoft image grand challenge on image retrieval (2015). http://press.liacs.nl/mmgrand/microsoft.pdf
44. Monmarché N (1999) On data clustering with artificial ants. In: AAAI-99 & GECCO-99 Workshop on Data Mining with Evolutionary Algorithms: Research Directions, Orlando, pp 23–26
45. Munkres J (1974) Topology, a first course. Prentice-Hall. http://books.google.co.in/books?id=LtEPAQAAMAAJ
46. Nash J (1951) Non-cooperative games. Ann Math 54(2):286–295. https://doi.org/10.2307/1969529
47. Parpinelli RS, Lopes HS, Freitas AA (2002) Data mining with an ant colony optimization algorithm. IEEE Trans Evol Comput 6(4):321–332
48. Patnaik S, Yang XS, Nakamatsu K (2017) Nature-inspired computing and optimization, vol 10. Springer, Cham
49. Powers DMW (2007) Evaluation: from precision, recall and F-factor to ROC, informedness, markedness & correlation. Technical report SIE-07-001, School of informatics and engineering, Flinders University, Adelaide
50. Ramakrishnan R, Gehrke J (2003) Database management systems, 3rd edn. McGraw-Hill, Inc., New York
51. Ramaswamy S, Rose K (2009) Towards optimal indexing for relevance feedback in large image databases. IEEE Trans Image Process 18:2780–2789
52. Rui Y, Huang T, Ortega M, Mehrotra S (1998) Relevance feedback: a power tool for interactive content-based image retrieval. IEEE Trans Circuits Syst Video Technol 8(5):644–655. https://doi.org/10.1109/76.718510
53. Schmitz P (2006) Inducing ontology from flickr tags. In: Collaborative Web Tagging Workshop at WWW'06, vol 50. Edinburgh, Scotland
54. Shelokar P, Jayaraman VK, Kulkarni BD (2004) An ant colony approach for clustering. Anal Chim Acta 509(2):187–195
55. Strauss K, Burger D (2014) What the future holds for solid-state memory. Computer 47(1):24–31
56. Su Z, Zhang H, Li S, Ma S (2003) Relevance feedback in content-based image retrieval: Bayesian framework, feature subspaces, and progressive learning. IEEE Trans Image Process 12(8):924–937
57. Sunet dataset (2013) http://www.cis.hut.fi/picsom/ftp.sunet.se
58. Tao D, Tang X (2004) Nonparametric discriminant analysis in relevance feedback for content-based image retrieval. In: Proceedings of the Pattern Recognition, 17th International Conference on (ICPR'04), vol 2. IEEE Computer Society, Washington, DC, pp 1013–1016
59. Tao D, Tang X (2004) Random sampling based SVM for relevance feedback image retrieval. In: Proceedings of the 2004 IEEE Computer Society Conference on Computer Vision and Pattern Recognition, CVPR'04. IEEE Computer Society, Washington, DC, pp 647–652
60. Torgerson W (1952) Multidimensional scaling: I. theory and method. Psychometrika 17:401–419. https://doi.org/10.1007/BF02288916

61. Tuncel E, Ferhatosmanoglu H, Rose K (2002) VQ-index: an index structure for similarity searching in multimedia databases. In: Proceedings of ACM Multimedia, Juan Les Pins pp 543–552
62. Vasconcelos N, Lippman A (2000) A unifying view of image similarity. In: 15th International Conference on Pattern Recognition. Proceedings, Barcelona, vol 1. IEEE, pp 38–41
63. Wang K, Yin Q, Wang W, Wu S, Wang L (2016) A comprehensive survey on cross-modal retrieval. arXiv preprint arXiv:1607.06215
64. Webb AR (1995) Multidimensional scaling by iterative majorization using radial basis functions. Pattern Recogn 28(5):753–759
65. Weber R, Schek HJ, Blott S (1998) A quantitative analysis and performance study for similarity-search methods in high-dimensional spaces. In: Proceedings of the 24rd International Conference on Very Large Data Bases, VLDB'98. Morgan Kaufmann Publishers Inc., San Francisco, pp 194–205
66. White DA, Jain R (1996) Similarity indexing with the SS-tree. In: Proceedings of the Twelfth International Conference on Data Engineering, ICDE'96. IEEE Computer Society, Washington, DC, pp 516–523
67. Wolfson HJ (1990) Model-based object recognition by geometric hashing. In: Proceedings of the First European Conference on Computer Vision, ECCV'90. Springer, New York, pp 526–536. http://portal.acm.org/citation.cfm?id=89081.89183
68. Xu X, Chen L, He P (2007) A novel ant clustering algorithm based on cellular automata. Web Intell Agent Syst 5(1):1–14. http://dl.acm.org/citation.cfm?id=1377757.1377758
69. Yin PY, Bhanu B, Chang KC, Dong A (2005) Integrating relevance feedback techniques for image retrieval using reinforcement learning. IEEE Trans Pattern Anal Mach Intell 27:1536–1551
70. Zhang L, Rui Y (2013) Image search—from thousands to billions in 20 years. ACM Trans Multimedia Comput Commun Appl 9(1s):36:1–36:20. https://doi.org/10.1145/2490823
71. Zhou XS, Huang TS (2003) Relevance feedback in image retrieval: a comprehensive review. Multimedia Syst 8(6):536–544. https://doi.org/10.1007/s00530-002-0070-3
72. Zhuo L, Cheng B, Zhang J (2014) A comparative study of dimensionality reduction methods for large-scale image retrieval. Neurocomputing 141:202–210. https://doi.org/10.1016/j.neucom.2014.03.014
73. Zhou W, Li H, Tian Q (2017) Recent advance in content-based image retrieval: a literature survey. arXiv preprint arXiv:1706.06064

Chapter 10
Filter-Based Feature Selection Methods Using Hill Climbing Approach

Saptarsi Goswami, Sanjay Chakraborty, Priyanka Guha, Arunabha Tarafdar, and Aman Kedia

10.1 Introduction

Feature selection is a critical component in most of the pattern recognition, machine learning, and data mining tasks. Apart from the obvious merits of sleeker models, better interpretability, and reduced data collection effort, it has been shown to improve performance over varied domains and tasks [1, 2]. Feature selection involves the process of finding the optimal solution, i.e., the optimal feature subset from a set of feasible solutions or feature subsets. The very nature of the feature selection problem makes it an ideal candidate for using a search-based optimization technique. Typically the possible feature sets are encoded as binary strings. As these solution spaces are discrete in nature (consisting of 0 and 1), this is also called a combinatorial optimization problem. Based on the associated objective function with the solution, the feature selection methods can be broadly classified as a filter method or a wrapper method. The filter method measures the merit or utility of a feature subset based on some statistical or information theoretic properties, whereas wrapper method measures the merit based on a particular machine learning model's performance. Another category of feature selection method is called embedded method. In these cases, loss function or the objective function has a component of feature selection included in them. Lasso [3] is an example of such a method. As filter methods are more generic in nature in this paper, the scope of the study has

S. Goswami (✉)
A.K. Choudhury School of IT, University of Calcutta, Kolkata, India

S. Chakraborty (✉)
Department of Information Technology, Techno India, Kolkata, India

P. Guha · A. Tarafdar · A. Kedia
Computer Science and Engineering, Institute of Engineering and Management, Kolkata, India

X. Li, K.-C. Wong (eds.), *Natural Computing for Unsupervised Learning*, Unsupervised and Semi-Supervised Learning,
https://doi.org/10.1007/978-3-319-98566-4_10

been confined to filter methods. A hybrid feature selection strategy was introduced for supervised and unsupervised classification using feature association map [4]. The selection of a good feature selection strategy depends on the characteristics of input datasets as well [5, 6].

A complete search using an exhaustive strategy creates a combinatorial explosion, even for a moderate size of the dataset. To overcome this problem on the one hand, various kinds of heuristic and evolutionary techniques are applied [7–9], and on the other hand, alternatives to search-based methods like clustering-based methods are explored [10–12]. Hill climbing has an intuitive appeal because of its simplicity and capacity of scaling well [13]. A criticism of the current work may be appropriateness of research in hill climbing, given its simplistic nature and long existence. Additionally it is also known for its tendency to get trapped in local minimum [14]. It is also found from literature that there are many metaheuristic techniques which provide better results than hill climbing [15]. Interestingly a search in Google Scholar reports 23,000 articles from 2013 onward related to hill climbing. Varied applications are seen in terms of Boolean circuit synthesis [16], university examination timetabling [17], photovoltaic systems [18], etc. In many cases, hill climbing is used in conjunction with other algorithms like clustering [19] and bird-mating optimization [20]. Newer variants like late acceptance hill climbing [15], which delays the comparison with the current best by some iteration or optimized hill climbing [21], are also being developed. These facts establish research communities continued interest in theoretical development as well as applications of hill climbing.

There can be some variants of hill climbing techniques based on the direction of the search, definition of the neighborhood, stochastic or deterministic approach in selecting the neighbor, etc. [22]. Feature selection using different variants of hill climbing has been employed in varieties of task [14, 21, 23–28], namely, natural language processing, image processing, spam classification, gait classification, etc. We felt that current literature does not provide answer to some key questions especially for choosing between the variants of hill climbing in the context of feature selection. In an earlier study [13], some of the questions related to the variants of the hill climbing were indeed answered with respect to propositional satisfiability (SAT) problem. Paper [29] is another study where authors have shown hill climbing to be effective for minimum grouping problem. However, these results and conclusions cannot be trivially extended to feature selection.

The questions that we raise and attempt to answer in this paper, in the context of feature selection, are as follows:

- How different is the result of a deterministic setup vis-a-vis a stochastic setup for neighbor selection strategy?
- While we change the search direction from random to forward or backward, how different is the result?
- What is the effect of exploring a 2-bit neighborhood instead of a 1-bit neighborhood?

- Genetic algorithm is more sophisticated with the selection, crossover, and mutation operators over hill climbing which has only mutation. GA depends on more parameters like population size, crossover probability, and mutation probability [30]. Consequently the models are less interpretable as compared to hill climbing. Does the improvement in performance justify this reduction of comprehensibility of the model?

The unique contribution of the paper is that we have tried answering all the above questions with an extensive empirical study conducted over 20 publicly available datasets. This paper also encloses a simplified taxonomy of the various hill climbing variants.

The rest of the paper is organized as follows. In Sect. 10.2, we have given a brief description of various hill climbing methods. In Sect. 10.3, we have outlined related works in feature selection using hill climbing. In Sect. 10.4, we have described the materials and methods employed in this study. In Sect. 10.5, we have presented the results and analysis. Section 10.6 contains the conclusion of the presented work.

10.2 Taxonomy of Hill Climbing Search

In this section, we have described the different variants of hill climbing search with the help of Fig. 10.1. Hill climbing is a local search-based optimization technique. It starts with an initial solution and gradually generates its neighboring successors. If the neighboring state produces better result than its predecessor, then the neighboring state is made as the current state. Though hill climbing is simple in its approach, it serves as an important building block to many sophisticated search techniques like beam search, tabu search, simulated annealing, and to some extent genetic algorithm. We have not included tabu search or beam search in this discussion. Though they can be argued to be the extension of hill climbing, they do not fall under traditional hill climbing categories.

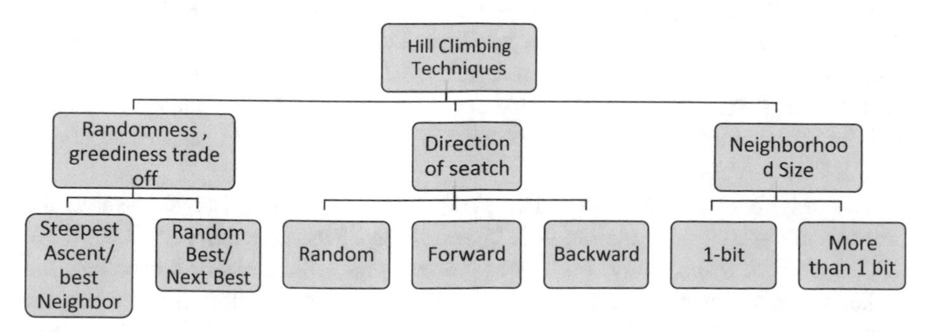

Fig. 10.1 Generalized taxonomy of hill climbing

We have divided the hill climbing techniques based on three characteristics. (i) How is the neighboring state selected from the current state? (ii) What is the direction of the search? (iii) What is the size of the neighborhood? One of the criticisms of hill climbing is its tendency to get trapped in local optima [14]. The first and third questions actually yield strategies to move out from the local optima. Another technique which is often used to escape local optima is the strategy of random restarts. However, as this can be applied to any of the basic methods, this has not been added as a parameter of classifying the hill climbing method. Iterated local search (ILS) [31] is a technique which starts by perturbing the local minimum and searching its neighborhood. This is also a technique to improve any local search, hence excluded from current study. Another important property of hill climbing is that it is an anytime algorithm, which can make it a good choice when there is a constraint in terms of availability of time [29]. Also, a variant-like late acceptance hill climbing [20] has not been discussed here to keep the current scope of the study more generalized.

Pseudocode of the most generalized version is enclosed below.

Algorithm Steepest ascent hill climbing

```
Input: Solution space (PS)
Output: Near optimal solution (OS)
OS← RandomSolution (PS)
PreviousFit← Cost(OS)
Neighbors←generateNeighbors(OS)
CurrentFit←best(Cost(Neighbors)
While (CurrentFit > PreviousFit)do
    PreviousFit← CurrentFit
Neighbors←generateNeighbors(OS)
     CurrentFit←best(Cost(Neighbors)
     OS←bestNeighbor(OS)
End Loop
```

Steepest Ascent Versus Random Best

Both these variants use greedy approach, but they are greedy in different ways. While steepest ascent deterministically chooses the best neighbor, random best neighbor chooses any arbitrary neighbor which does not reduce the objective function value. Random best is better positioned to not get trapped in a local optima. The difference is further illustrated using Fig. 10.2. The problem in the example is assumed to be maximization problem. Let us assume the size of the problem is 4 and the initial solution is 1001. The numbers beside the subset indicate the fitness value of the solution.

The neighbors of 1001 are 0001, 1101, 1011, and 1000. Let us assume the fitness value of the first solution is 0.75 and those of the neighbors are 0.70, 0.76, 0.80, and 0.65, respectively. As it is assumed to be a maximization problem, the higher the value of the objective function, the better the solution. Steepest ascent deterministically chooses the solution 1011, whereas random best may choose any one of 1101 and 1011. Instead of equal probability, it may be made to vary with

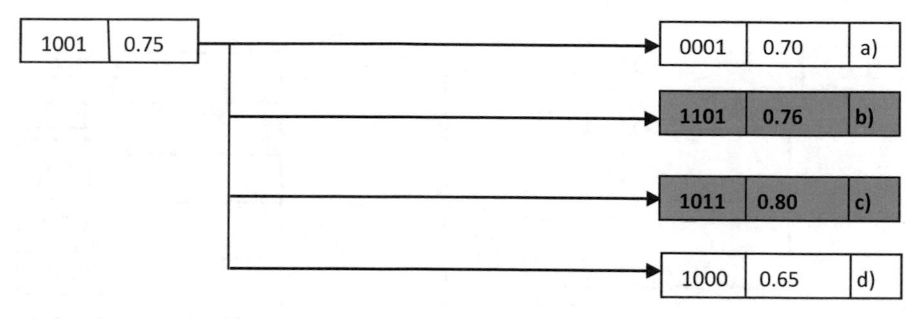

Fig. 10.2 Neighbor selection strategy

the gradient of the candidate solution. The random best strategy is also known as random mutation hill climbing [32].

Direction-Based Hill Climbing

This can be classified in three categories. It may be noted for all three of the categories, any of steepest ascent and random best strategy can be adopted. Some other variants like bidirectional hill climbing can be seen from literature [28] which can allow attributes to be deleted and added. However, random direction also allows going in both directions.

- *Forward Hill Climbing:* It starts from the binary string with all 0 s. Then subsequently number of 1 s in the candidate solutions increases gradually by 1. So a possible solution path may be 0000-0001-0101-1101.
- *Backward Hill Climbing:* It starts with the binary string with all 1 s. Then subsequently number of 1 s in the candidate solution decreases gradually by 1. So a possible solution path may be 1111-1101-1001. Backward is better poisoned to account for feature interaction than forward.
- *Random Hill Climbing:* Random hill climbing is a direct search-based technique which randomly selects its neighbor of a best-so-far solution and accepts a neighbor if it is found better or equal to it. In our proposed method, there are no such variations of direction-based hill climbing. The above discussion is contributed to represent a basic understanding of the techniques which can be used further to produce a better modified result.

Neighborhood Size

The general strategy is using a 1-bit neighborhood, i.e., all possible solutions where only 1 bit is mutated from the current solution form the neighborhood. For a 2-bit neighborhood, the same concept is extended and 2 bits are mutated randomly. The difference in possible solution path is illustrated in Fig. 10.3.

While 1-bit strategy produces n neighbors of the current solution, a "*k*"-bit strategy will produce $\binom{n}{k}$ neighbors.

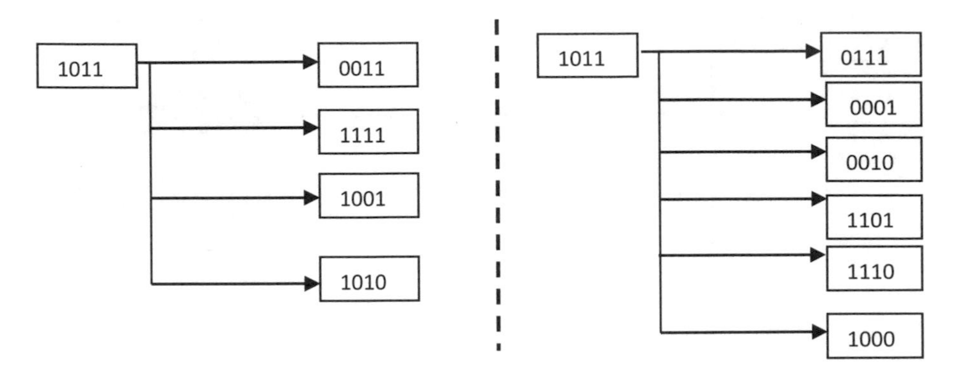

Fig. 10.3 1-bit versus 2-bit neighborhood

10.3 Hill Climbing in Feature Selection

Paper [23] uses hill climbing with correlation-based feature selection (CFS) [33] as the fitness or objective function. Hill climbing is used in both forward search and backward search setup. Results were comparable with wrappers and all features and it was much faster. Paper [24] illustrated applicability of feature selection for credit scoring and used forward hill climbing as one of the search strategies. It is observed that the said strategy brings good result as compared to other expensive strategies in much lesser time. Paper [14] used hill climbing for feature selection with a random start, and then ten iterations of the same have been used. The problem domain was that of classification of emails into spam and non-spam. It was used in a wrapper model with random best neighbor selection strategy. Hill climbing (HC) improved results over LDA; however simulated annealing and threshold accepting performed better than HC. Paper [21] uses a hill climbing search for handwritten character recognition. It uses a wrapper-based approach where the inducted model is a MLP neural network. The features have been prioritized based on their error rates, and a level has been used for discretization. Paper [25] discussed hill climbing with respect to gait classification. The induction algorithm used is SVM and the metric that was optimized was receiver operating characteristics (ROC). Paper [26] have used random mutation hill climbing for an image classification task. The experiments have been conducted in a wrapper fashion with a couple of successful improvements: (a) A number of bits to be mutated are decreased gradually. (b) The fitness function is refined to include a factor on number of features remaining apart from the misclassification rate. In paper [27] authors have worked with the problem of paraphrase generation which is an important natural language processing task. They have employed feature selection in a wrapper approach. For feature selection they have used both forward hill climbing (FHC) and forward beam search (FBS) (beam width = 10). From experimental result it was observed that FHC was equally effective to FBS. In paper [28], authors have applied five variants of hill climbing-based feature selection for two learning tasks taken from a calendar scheduling domain. It may be summarized that feature selection using hill climbing has been used in many applications. Different variants have been applied in different cases in somewhat ad hoc fashion. One other observation is most of them have used a wrapper method of feature selection. The summary is also presented in Table 10.1.

10.4 Materials and Methods

In this section we discuss about the details of the simulation experiment. The datasets have been taken from the public dataset repository of UCI [34] and Keel dataset repository [35].

10.4.1 Datasets

In Table 10.2, the list of the datasets as used in the experiment is mentioned. The table contains a brief description of the dataset, along with number of attributes/features, number of classes, and number of instances/records or observations. The datasets are from various domains like image processing, medical, and agriculture to name a few. Some datasets like seeds or ionosphere have two to three classes, whereas some datasets like digits have ten classes. In terms of attributes, many are small, where some have more than 50, 100, or 200 features. Consequently it can be claimed that the test bed has good diversity. Hence the inferences drawn may be generalized to a fair extent.

10.4.2 Computational Environment, Libraries, and Parameters

"R" has been used as the computational environment [36], and some of the libraries that have been used are "GA" [37], "miscFuncs" [38], "entropy" [39], etc. In hardware specification, we are running these algorithms, and the results are analyzed on Windows 7, 4 GB RAM, and Core I3 processor architectures. The other parameters and assumptions used are as follows:

- The crossover probability is taken as 0.7. Natural % of couples who had at least one child is 70.85% [40].

  Mutation probability is taken as 0.05. If these probabilities are too low, then the algorithm will not cover the search space much. If they are too high, then good candidate solutions will be perturbed and pushed into worse solutions. However, a rough measure is the rate of improvement in the fitness function.

Table 10.1 Summary of literature study

Reference	Hill climbing type	Filter/wrapper	Application domain	Remarks
Hall and Smith [23]	Backward and forward	Filter	Generic	CFS was used as the objective function
Liu and Schumann [24]	Forward	Both	Credit scoring	–
Wang et al. [14]	Multiple restart and random neighbor strategy	Wrapper	Spam detection in email	–
Nunes et al. [21]	Some modification over original hill climbing using priority score	Wrapper	Handwritten character recognition	–
Begg et al. [25]	Forward hill climbing	Wrapper	Gait classification	ROC was used as the fitness function
Farmer 2004 [26]	Random mutation with some improvement	Wrapper	Image classification	Classification rate as used as objective function
Malakasiotis et al. [27]	Forward hill climbing with beam search	Wrapper	Paraphrase learning for natural language processing	Accuracy, precision, recall, and F-measure was used
Freitag [28]	Forward, backward with flexibility of going back	Wrapper	Calendar scheduling problem	–
Proposed method using hill climbing approach	Steepest ascent neighbor, 1-bit neighborhood, and random search	Filter	Efficiently handle high-dimensional datasets by feature subset selection	CFS is used as the objective function

Table 10.2 Dataset details

Name	Description	Number of attributes	Number of classes	Number of instances
Ionosphere	Classification of radar returns from the ionosphere	34	2	351
Biodegradation	Relationships between chemical structure and biodegradation of molecules	41	2	1055
Seeds	Measurements of geometrical properties of kernels belonging to three different varieties of wheat	7	3	210
Magic	Data are MC generated to simulate registration of high-energy gamma particles in an atmospheric Cherenkov telescope	11	2	19,020
Breast tissue	Dataset with electrical impedance measurements of freshly excised tissue samples from the breast	10	6	106
Appendicitis	The data represents 7 medical measures taken over 106 patients on which the class label represents if the patient has appendicitis (class label 1) or not (class label 0)	7	2	106
New thyroid	This dataset is one of the several databases about thyroid available at the UCI repository. The task is to detect if a given patient is normal (1) or suffers from hyperthyroidism (2) or hypothyroidism (3)	5	3	215
Heart	This dataset is a heart disease database	13	2	270
Wine	Using chemical analysis determine the origin of wines	13	3	178
Bupa	This dataset analyzes some liver disorders that might arise from excessive alcohol consumption	6	2	345
Cardiotocography	2126 fatal cardiotocograms (CTGs) were automatically processed and the respective diagnostic features measured	35	10	2126
Breast cancer	Original Wisconsin Breast Cancer Database	10	2	699

(continued)

Table 10.2 (continued)

Name	Description	Number of attributes	Number of classes	Number of instances
Hepatitis	Hepatitis dataset	20	2	155
Cleveland	This dataset is a part of the heart disease dataset	13	5	297
E. coli	This data contains protein localization sites	7	6	336
Parkinson	Oxford Parkinson's Disease Detection Dataset	22	2	197
Optdgt	Optical recognition of handwritten digits dataset	64	10	5620
Digit	1593 handwritten digits from around 80 persons were scanned, stretched in a rectangular box 16x16 in a gray scale of 256 values	256	10	1593
Sonar	The task is to train a network to discriminate between sonar signals bounced off a metal cylinder and those bounced off a roughly cylindrical rock	60	2	208
LSVT	126 samples from 14 participants, 309 features. Aim: assess whether voice rehabilitation treatment leads to phonations considered "acceptable" or "unacceptable"	309	2	126

- The population size used for GA is 10 times the number of features for that particular dataset. GA practitioners often use high population sizes to find better solution. If your population size is too small, the improvement per iteration in the fitness function will be low.
- Number of iterations the search process is executed for GA and random best neighbor hill climbing is 200 iterations. This number of iterations is related to improvement in the fitness function. A fitness function usually shows major improvement in early iterations and then asymptotically approaches an optimum.
- The neighborhood sizes explored are 1-bit and 2-bit, respectively.
- For comparing, results of the search techniques employed in feature selection classification accuracy are used.
- The classification accuracy reported is the arithmetic mean of 100 iterations.
- We have used the validation set approach for adjusting the training model bias.
- For classification model, recursive partitioning and regression trees have been used [41].
- In all the comparisons in the next section, hill climbing with steepest ascent neighbor, 1-bit neighborhood, and random direction of search has been considered as the baseline case.

- While comparing genetic algorithm with hill climbing, the best result of the available variants has been compared with GA. Both the GA and the hill climbing methods run in the same environment (above specified) in this work.
- We have run our experiments using a widely cited objective function. The fitness function used is correlation-based feature selection (CFS) [33].

CFS is given as

$$S_{\mathrm{CFS}} = \frac{k\overline{r}_{\mathrm{zc}}}{\sqrt{k + k * (k-1) * \overline{r}_{ij}}} \tag{10.1}$$

where S_{CFS} is a measure indicating quality of a feature subset, r_{zc} is the average of correlation between the features and the target variable, and r_{ij} is the average inter-correlation between the components. "k" is the cardinality of the selected feature subset. Most of the objective functions use the concept of feature relevance, which can be measured using correlation coefficient, Fisher's score, mutual information, and message digest length (MDL) [33].

10.5 Results and Analysis

In this section, we discuss firstly about the effect of randomness over the gradient descent setup. Next, we discuss about effect of direction in hill climbing search and effect of neighborhood. Finally we enclose a comparison with GA.

10.5.1 Random Neighbor Versus Steepest Ascent

The results in terms of classification accuracy are enclosed in Table 10.3. In the table, wherever the random best neighbor has achieved a better or equivalent result is marked in bold.

A comparison in terms of reduction and accuracy is given in Fig. 10.4.

It can be observed from Table 10.3 and Fig. 10.4 that:

- In terms of classification accuracy, they perform equivalently, while the best neighbor strategy command a 5% more reduction in number of features.
- However, a detailed observation reveals that on a head-to-head basis, best neighbor strategy lags with four wins, eight draws, and eight losses.
- Most interestingly for both the datasets (LSVT, digit) with more than 250 features, the random neighbor strategy yields much better result as compared to best neighbor strategy.

Table 10.3 Random best neighbor versus best neighbor

	Best neighbor		Random best neighbor	
Datasets	Accuracy	Reduction	Accuracy	Reduction
Ionosphere	0.775	0.970	0.755	0.848
Biodegradation	0.729	0.951	**0.767**	0.927
Seeds	0.857	0.286	**0.857**	0.286
Magic	0.731	0.800	0.725	0.800
Breast tissue	0.375	0.778	**0.375**	0.778
Appendicitis	0.841	0.000	**0.843**	0.000
New thyroid	0.862	0.600	**0.862**	0.600
Heart	0.778	0.692	**0.779**	0.692
Wine	0.843	0.846	0.815	0.846
Bupa	0.573	0.667	**0.573**	0.667
Cardiotocography	0.900	0.879	**0.901**	0.848
Parkinson's	0.763	0.909	0.762	0.909
Breast cancer	0.938	0.444	**0.938**	0.444
Hepatitis	0.809	0.895	**0.809**	0.895
Cleveland	0.621	0.692	**0.622**	0.692
E. coli	0.792	0.125	**0.792**	0.125
Sonar	0.671	0.650	**0.676**	0.417
Optdgt	0.296	0.887	**0.296**	0.742
LSVT	0.697	0.994	**0.755**	0.942
Digits	0.257	0.973	**0.282**	0.828

10.5.2 The Direction of the Search

Classification accuracies as obtained by the three search directions are presented below in Table 10.4.

The comparison in terms of mean accuracy and mean reduction achieved is depicted in Fig. 10.5.

It can be observed from Table 10.4 and Fig. 10.5 that:

- Their performance seems to be very similar to each other in terms of classification accuracy.
- The p-values of pairwise t-test between random and forward, random and backward, and finally forward and backward are 0.10, 0.29, and 0.43, respectively, for a degree of freedom of 19. This indicates that the marginal deviations that the methods exhibit do not have any statistical significance.
- One interesting fact that comes out is that on an average backward and random search direction achieves a better performance as far as reduction is concerned.

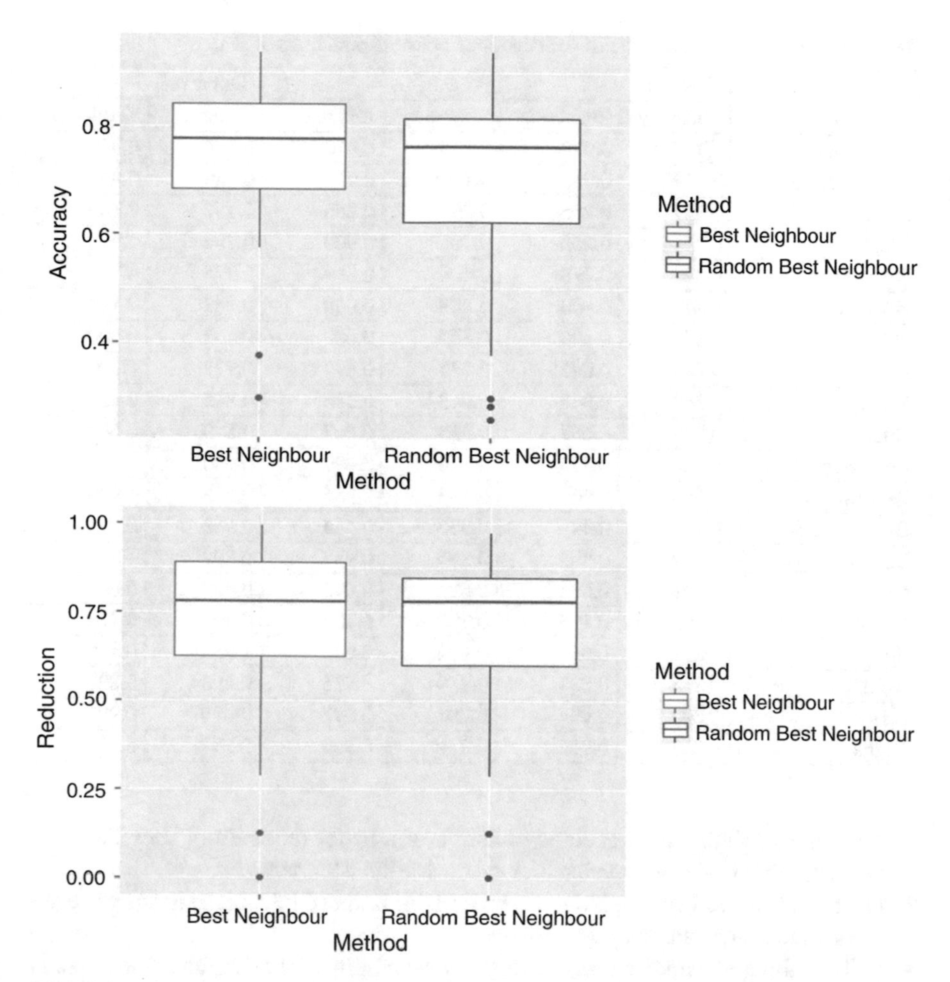

Fig. 10.4 Comparing mean accuracy and reduction rate of best neighbor vs random neighbor strategy

10.5.3 Size of Neighborhood

In this paper, we have kept the scope to be 1-bit and 2-bit, respectively. The classification accuracies as obtained from the two methods are reported in Table 10.5. The results where 1-bit has at least drawn with 2-bit are marked in bold.

A comparison in terms of accuracy and reduction is shown in Fig. 10.6. It may however be noted that the results of digit and LSVT have not been included as the search processes run for days for the 2-bit neighborhood.

However, it can be observed from Table 10.5 and Fig. 10.6 that:

Table 10.4 Classification accuracies with the three possible directions of search

Datasets	Random		Forward		Backward	
	Accuracy	Reduction	Accuracy	Reduction	Accuracy	Reduction
Ionosphere	0.775	0.970	0.726	0.970	0.774	0.939
Biodegradation	0.729	0.951	0.706	0.976	0.669	0.951
Seeds	0.857	0.286	0.857	0.286	0.857	0.286
Magic	0.731	0.800	0.706	0.900	0.706	0.900
Breast tissue	0.375	0.778	0.404	0.889	0.375	0.778
Appendicitis	0.841	0.000	0.842	0.000	0.842	0.143
New thyroid	0.862	0.600	0.862	0.600	0.862	0.600
Heart	0.778	0.692	0.777	0.692	0.777	0.692
Wine	0.843	0.846	0.843	0.385	0.845	0.385
Bupa	0.573	0.667	0.533	0.667	0.573	0.667
Cardiotocography	0.900	0.879	0.900	0.879	0.900	0.879
Parkinson's	0.763	0.909	0.763	0.955	0.763	0.909
Breast cancer	0.938	0.444	0.938	0.444	0.938	0.444
Hepatitis	0.809	0.895	0.809	0.947	0.809	0.895
Cleveland	0.621	0.692	0.622	0.692	0.623	0.692
E. coli	0.792	0.125	0.792	0.125	0.792	0.125
Sonar	0.671	0.650	0.672	0.650	0.682	0.550
Optdgt	0.296	0.887	0.289	0.935	0.296	0.823
LSVT	0.697	0.994	0.703	0.997	0.696	0.997
Digits	0.257	0.973	0.220	0.969	0.257	0.973

- 1-bit enjoys slight better accuracy in spite of a better reduction. (They have the same number of outlying values for classification accuracy.)
- On a head-to-head comparison, it reveals 1-bit is ahead of 2-bit in terms of seven wins, nine draws, and only two losses.
- Hill climbing apparently loses its edge of scalability if the neighborhood size is increased as observed for the datasets which had more than 100 features.

To summarize about the variants about hill climbing in filter-based feature selection:

- Random best strategy seems to work better than that of steepest ascent. In 80% of the datasets, random best strategy produced equivalent or better result.
- The direction of search does not seem to affect the results in statistically significant way.
- A 1-bit strategy produces better result than 2-bit strategy and produces equivalent or better result in 90% of the datasets. Also, as the dataset size grows, the 2-bit strategy becomes quite computationally expensive.

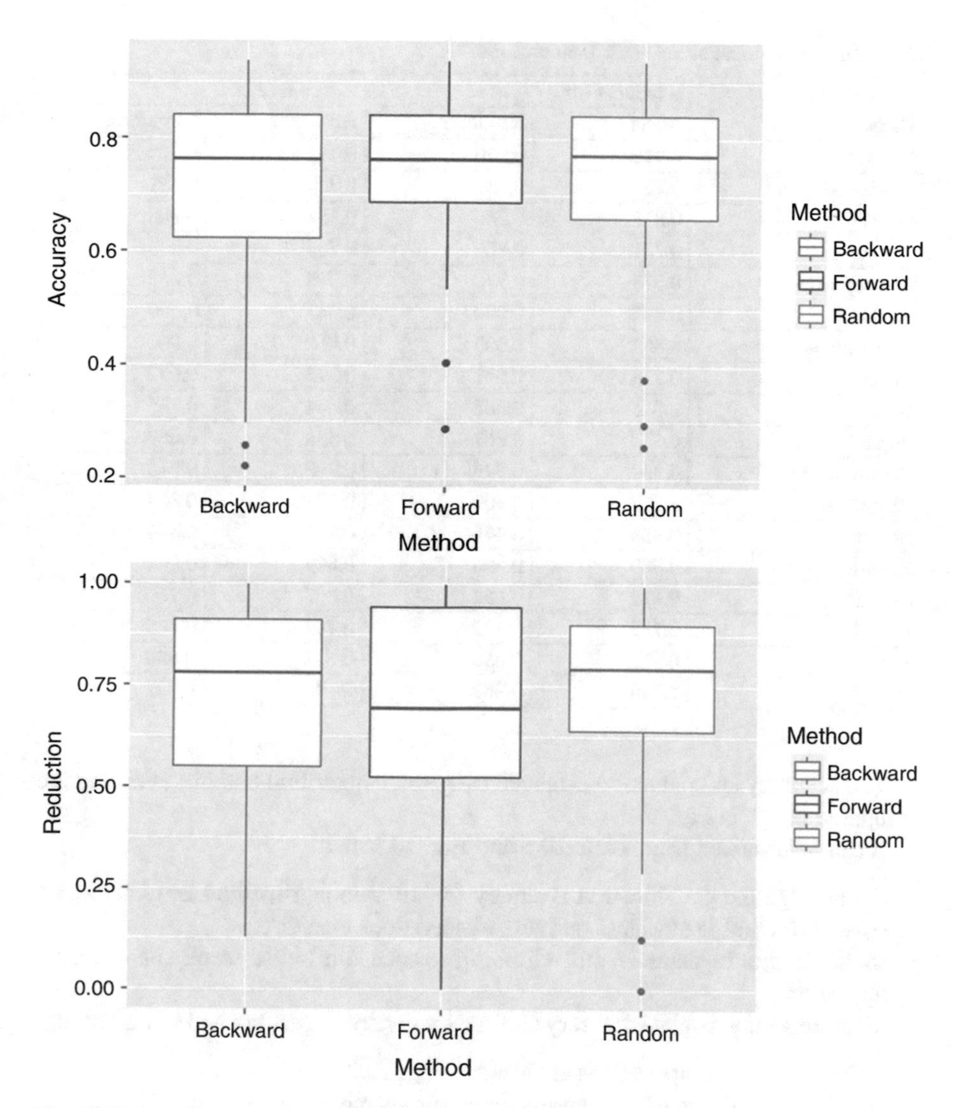

Fig. 10.5 Comparing mean accuracy and reduction rate based on search direction

10.5.4 Hill Climbing Versus Genetic Algorithm

In this section, we present a comparison of hill climbing with the genetic algorithm. Though there are some criticisms about GA, it is one of the most popular algorithms and is quite sophisticated in its approach as compared to hill climbing. In Table 10.6, we have presented the result. The datasets where hill climbing has produced better result have been marked in bold.

Table 10.5 Classification accuracy 1-bit vs 2-bit

	Random (1 bit)		Random (2 bit)	
Datasets	Accuracy	Reduction	Accuracy	Reduction
Ionosphere	**0.775**	0.970	0.755	0.879
Biodegradation	**0.729**	0.951	0.729	0.951
Seeds	**0.857**	0.286	0.857	0.286
Magic	**0.731**	0.800	0.706	0.800
Breast tissue	**0.375**	0.778	0.375	0.667
Appendicitis	0.841	0.000	0.842	0.000
New thyroid	**0.862**	0.600	0.862	0.600
Heart	**0.775**	0.692	0.775	0.692
Wine	**0.846**	0.385	0.844	0.385
Bupa	**0.573**	0.667	0.533	0.667
Cardiotocography	**0.900**	0.879	0.900	0.848
Parkinson's	**0.763**	0.909	0.754	0.864
Breast cancer	**0.938**	0.444	0.937	0.444
Hepatitis	**0.809**	0.895	0.809	0.842
Cleveland	**0.621**	0.692	0.607	0.769
E. coli	**0.792**	0.125	0.792	0.125
Sonar	0.671	0.650	0.672	0.650
Optdgt	**0.296**	0.887	0.296	0.887

The accuracy and reduction achieved by genetic algorithm and hill climbing are compared below in Fig. 10.7.

It can be observed from Table 10.6 and Fig. 10.7 that:

- On average the classification accuracy of hill climbing method is 1.7% better. (Even hill climbing has less outlying observations than GA.)
- In 12 of the 20 datasets, hill climbing produces a better result than genetic algorithm.
- Mean reduction achieved is very similar for genetic algorithm and hill climbing.

A dataset-wise comparison is enclosed in Fig. 10.8.

Hill climbing is one of the baseline heuristic search methods, which is simple to understand, depends on minimal parameters, and is highly scalable. The objective of the paper was to choose the best variant of hill climbing for the task of feature selection. We also wanted to compare if this such selected best variant is competitive with more sophisticated heuristic algorithm like genetic algorithm, which is again a benchmark algorithm and has given better or competitive result on a diverse set of optimization problems. To compare them a test bed of 18 benchmark datasets have been chosen from publicly available datasets. The datasets have good variety in terms of size and application domain and hence the results can be generalized.

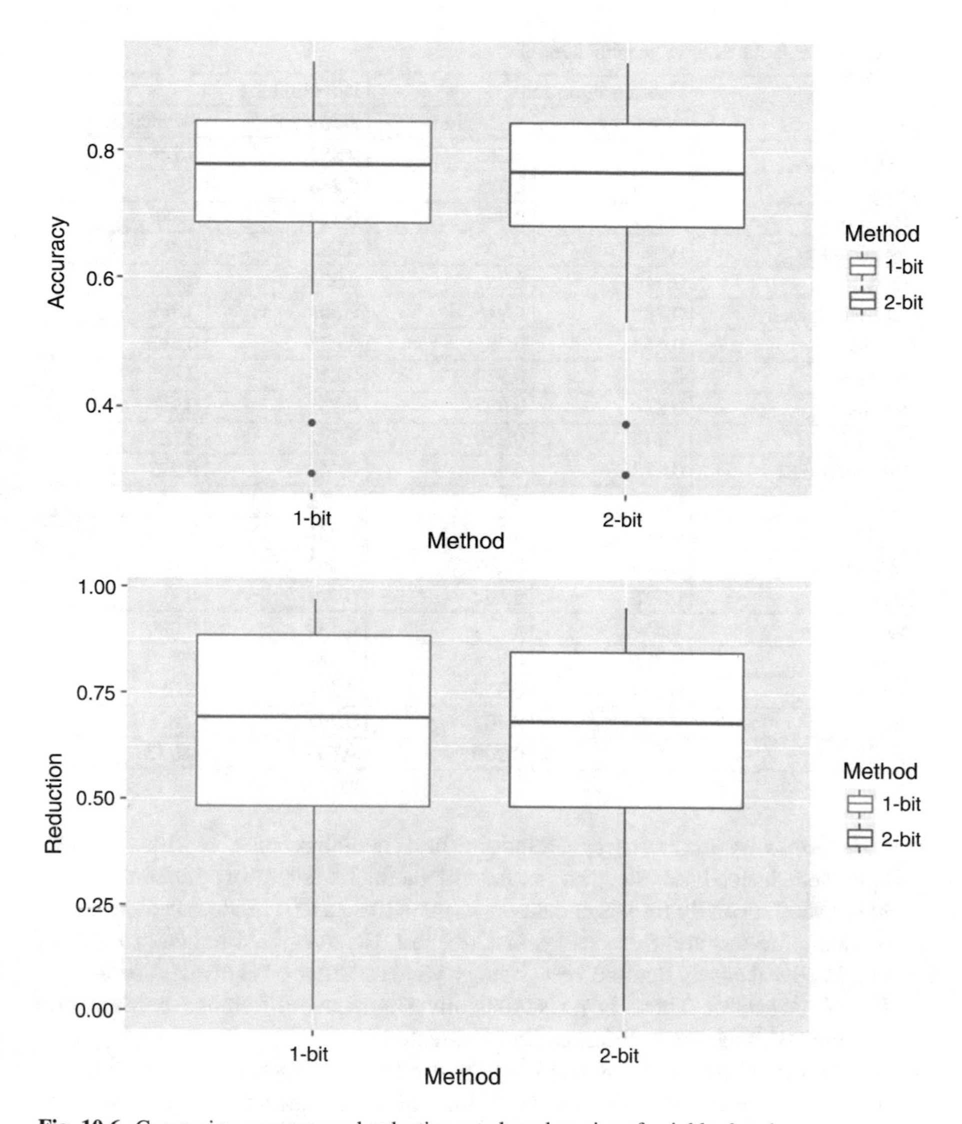

Fig. 10.6 Comparing accuracy and reduction rate based on size of neighborhood

10.6 Conclusion

Hill climbing is the simplest of the heuristic-based search techniques, and it is quite scalable. There are many variants of the hill climbing method in terms of random versus deterministic setup, direction of search, and size of neighborhood. This paper attempts to answer some of the key questions related to hill climbing for the domain of filter-based feature selection. The findings of our study over 20 publicly available datasets are as follows:

Table 10.6 Hill climbing vs genetic algorithm

Dataset	Genetic algorithm		Hill climbing	
	Accuracy	Reduction	Accuracy	Reduction
Seeds	0.873	0.286	0.857	0.286
Wine	0.796	0.385	**0.846**	0.385
Breast tissue	0.344	0.889	**0.404**	0.889
Breast cancer	0.947	0.500	0.938	0.444
Bupa	0.547	0.857	**0.573**	0.667
Cardiotocography	0.788	0.916	**0.901**	0.848
Cleveland	0.672	0.785	0.623	0.692
Hepatitis	0.830	0.950	0.809	0.895
Parkinson's	0.881	0.957	0.763	0.909
E. coli	0.743	0.250	**0.792**	0.125
Biodegradation	0.678	0.976	**0.767**	0.927
Ionosphere	0.745	0.333	**0.775**	0.970
Magic	0.712	0.900	**0.731**	0.800
New thyroid	0.846	0.600	**0.862**	0.600
Heart	0.716	0.692	**0.779**	0.692
Appendicitis	0.867	0.714	0.843	0.000
Sonar	0.736	0.783	0.682	0.550
Optdgt	0.244	0.935	**0.296**	0.887
Digits	0.291	0.707	0.292	0.878
LSVT	0.684	0.700	**0.755**	0.942

- *Neighbor selection strategy:* Random best neighbor may be the baseline approach instead of steepest ascent method. Its superior performance is manifested specially for larger datasets. Out of 20 datasets, random best produces 8 wins, 4 draws, and 8 losses against steepest ascent or best neighbor method, i.e., in 80% datasets random best strategy yields a better or equivalent result.
- *Search direction:* There is no statistically significant difference between the methods with the three different search directions.
- *Neighborhood size:* Moving from 1-bit to 2-bit neighborhood does not bring any general improvement in the results. In fact for larger datasets, the process tends to get much more computationally expensive. Consequently it loses the edge of scalability. 1-bit method dominates over 2-bit with seven wins, nine draws, and only two losses, i.e., in 90% of the datasets, 1-bit method produces a better or equivalent result.
- *Performance comparison with genetic algorithm:* Though GA is a much more sophisticated technique as compared to hill climbing, on our test bed, it was observed that hill climbing produced a better classification accuracy by 1.7%. Out of 20 datasets, hill climbing gives better result in 12 of the datasets.

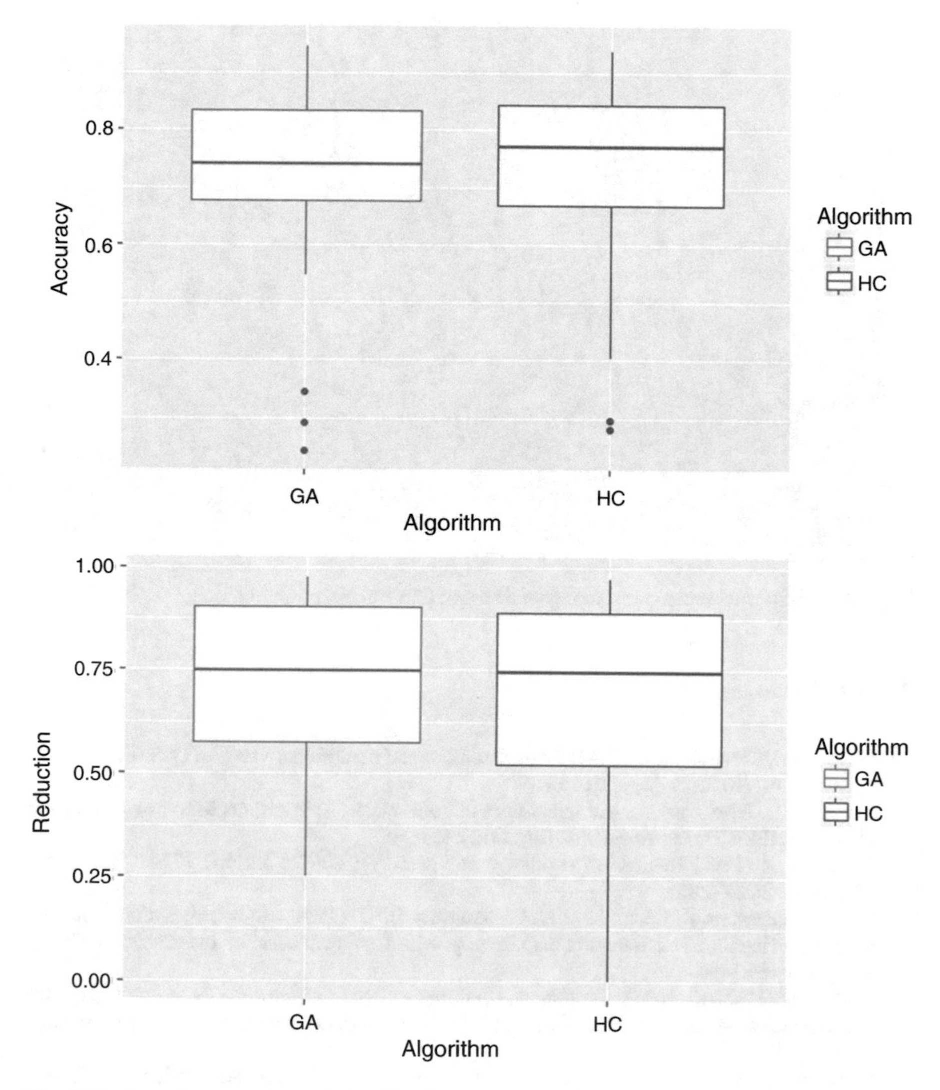

Fig. 10.7 Overall comparison of classification accuracy and reduction of genetic algorithm and hill climbing

Finally to conclude, methods like hill climbing are of interest given its property of easy scalability. This is particularly important as the sizes of datasets are exponentially increasing. If different parameters of hill climbing can be leveraged, it can potentially match or better the performance of more sophisticated algorithm like genetic algorithm.

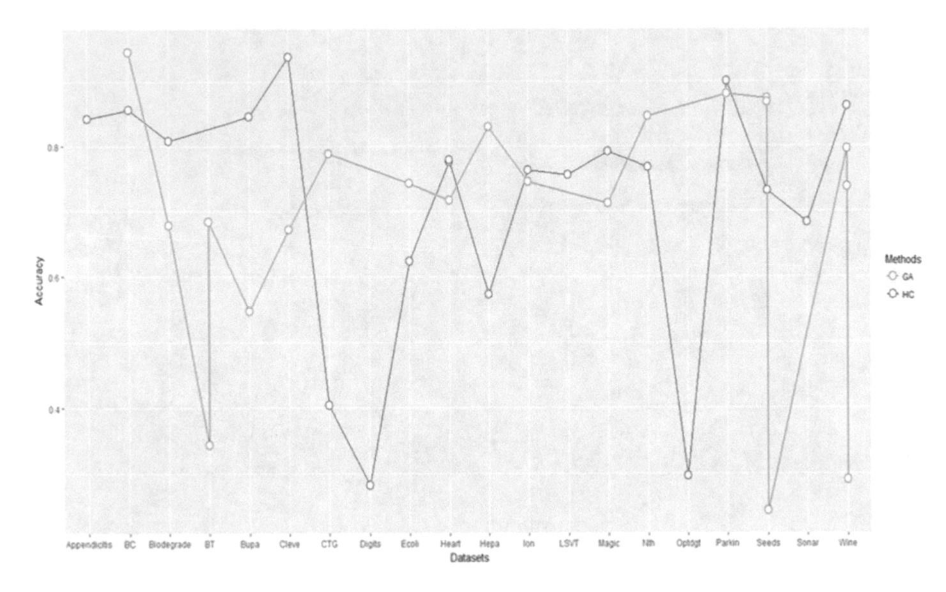

Fig. 10.8 Dataset-wise comparison of genetic algorithm and hill climbing

References

1. Goswami S, Chakrabarti A (2014) Feature selection: a practitioner view. IJITCS 6(11):66–77. https://doi.org/10.5815/ijitcs.2014.11.10
2. Liu H, Yu L (2005 Apr) Toward integrating feature selection algorithms for classification and clustering. IEEE Trans Knowl Data Eng 17(4):491–502
3. Tibshirani R (1996) Regression shrinkage and selection via the lasso. J R Stat Soc Ser B Methodol 58:267–288
4. Das AK, Goswami S, Chakrabarti A, Chakraborty B (2017) A new hybrid feature selection approach using feature association map for supervised and unsupervised classification. Expert Syst Appl 88:81–94
5. Goswami S, Das AK, Guha P, Tarafdar A, Chakraborty S, Chakrabarti A, Chakraborty B (2017) An approach of feature selection using graph-theoretic heuristic and hill climbing. Pattern Anal Applic:1–17
6. Goswami S, Chakrabarti A, Chakraborty B (2016) A proposal for recommendation of feature selection algorithm based on data set characteristics. J UCS 22(6):760–781
7. Goswami S, Saha S, Chakravorty S, Chakrabarti A, Chakraborty B (2015) A new evaluation measure for feature subset selection with genetic algorithm. Int J Intell Syst Appl MECS 7(10):28
8. Gheyas IA, Smith LS (2010) Feature subset selection in large dimensionality domains. Pattern Recogn 43(1):5–13
9. De La Iglesia B (2013) Evolutionary computation for feature selection in classification problems. Wiley Interdiscip Rev Data Min Knowl Disc 3(6):381–407
10. Goswami S, Das AK, Chakrabarti A, Chakraborty B (2017) A feature cluster taxonomy based feature selection technique. Expert Syst Appl 79:76–89
11. Goswami S, Chakraborty S, Saha HN (2017) An univariate feature elimination strategy for clustering based on metafeatures. Int J Intell Syst Appl 9(10):20

12. Goswami S, Chakrabarti A, Chakraborty B (2017) An efficient feature selection technique for clustering based on a new measure of feature importance. J Intell Fuzzy Syst 32(6):3847–3858
13. Gent IP, Walsh T (1993) Towards an understanding of hill-climbing procedures for SAT. In: AAAI, vol 93, pp 28–33
14. Wang R, Youssef AM, Elhakeem AK (2006) On some feature selection strategies for spam filter design. In: Electrical and computer engineering, 2006. CCECE'06, Canadian Conference on 2006 May. IEEE, pp 2186–2189
15. Burke EK, Bykov Y (2008) A late acceptance strategy in hill-climbing for exam timetabling problems. PATAT 2008 Conference, Montreal
16. Lang KJ (2016) Hill climbing beats genetic search on a boolean circuit synthesis problem of koza's. In: Proceedings of the twelfth international conference on machine learning 2016 Jan 22, pp 340–343
17. Bykov Y, Petrovic S (2016) A step counting hill climbing algorithm applied to university examination timetabling. J Schedul:1–4
18. Seyedmahmoudian M, Horan B, Rahmani R, Maung Than Oo A, Stojcevski A (2016) Efficient photovoltaic system maximum power point tracking using a new technique. Energies 9(3):147
19. Saichandana B, Srinivas K, Kumar RK (2014) Clustering algorithm combined with hill climbing for classification of remote sensing image. Int J Electr Comput Eng 4(6):923–930
20. Ou TC, Su WF, Liu XZ, Huang SJ, Tai TY (2016) A modified bird-mating optimization with hill-climbing for connection decisions of transformers. Energies 9(9):671
21. Nunes CM, Britto AS, Kaestner CA, Sabourin R (2004) An optimized hill climbing algorithm for feature subset selection: Evaluation on handwritten character recognition. In: Frontiers in handwriting recognition, 2004. IWFHR-9 2004. Ninth international workshop on 2004 Oct 26. IEEE, pp 365–370
22. Gelbart D, Morgan N, Tsymbal A (2009) Hill-climbing feature selection for multi-stream ASR. In: INTERSPEECH 2009, pp 2967–2970
23. Hall MA, Smith LA (1997) Feature subset selection: a correlation based filter approach. In: International conference on neural information processing and intelligent information systems, pp 855–858
24. Liu Y, Schumann M (2005) Data mining feature selection for credit scoring models. J Oper Res Soc 56(9):1099–1108
25. Begg RK, Palaniswami M, Owen B (2005) Support vector machines for automated gait classification. IEEE Trans Biomed Eng 52(5):828–838
26. Farmer ME, Bapna S, Jain AK (2004) Large scale feature selection using modified random mutation hill climbing. In: Pattern recognition, 2004. ICPR 2004. Proceedings of the 17th international conference on 2004 Aug 23, vol 2. IEEE, pp 287–290
27. Malakasiotis P (2009) Paraphrase recognition using machine learning to combine similarity measures. In: Proceedings of the ACL-IJCNLP 2009 student research workshop 2009 Aug 4. Association for Computational Linguistics, pp 27–35
28. Caruana R, Freitag D (1994) Greedy Attribute Selection. In: ICML, pp 28–36
29. Lewis R (2009) A general-purpose hill-climbing method for order independent minimum grouping problems: A case study in graph colouring and bin packing. Comput Oper Res 36(7):2295–2310
30. Mitchell M, Holland JH, Forrest S (2014) Relative building-block fitness and the building block hypothesis. D. Whitley. Found Genet Algorithms 2:109–126
31. Lourenço HR, Martin OC, Stützle T (2003) Iterated local search. In: Handbook of metaheuristics. Springer, Boston, pp 320–353
32. Mitchell M, Holland JH When will a genetic algorithm outperform hill-climbing?
33. Hall MA Correlation-based feature selection for machine learning. Doctoral dissertation, The University of Waikato
34. Lichman M (2013) UCI machine learning repository [http://archive.ics.uci.edu/ml]. University of California, School of Information and Computer Science, Irvine

35. Alcalá-Fdez J, Fernandez A, Luengo J, Derrac J, García S, Sánchez L, Herrera F (2011) KEEL Data-Mining Software Tool: Data Set Repository, Integration of Algorithms and Experimental Analysis Framework. J Mult Valued Log Soft Comput 17(2-3):255–287
36. R Core Team (2013) R: A language and environment for statistical computing. R Foundation for Statistical Computing, Vienna, Austria. ISBN 3-900051-07-0, URL http://www.R-project.org/
37. Luca Scrucca (2013) GA: A Package for Genetic Algorithms in R. Journal of Statistical Software, 53(4), 1–37. URL,http://www.jstatsoft.org/v53/i04/
38. Taylor BM (2013) miscFuncs: miscellaneous useful functions. R package version 1.2-4. http://CRAN.R-project.org/package=miscFuncs
39. Hausser J, Strimmer K (2012) entropy: entropy and mutual information estimation. R package version 1.1.7 http://CRAN.R-project.org/package=entropy
40. Gutowski MW (2005) Biology, physics, small worlds and genetic algorithms. In: Shannon S (ed) Leading edge computer science research. Nova Science Publishers Inc, Hauppage, pp 165–218
41. Therneau T, Atkinson B, Ripley B (2012) rpart: recursive partitioning. R package version 4.1-0

Part IV
Others

Chapter 11
Social Networking Awareness in Indian Higher Education

Shamdeep Brar, Tomayess Issa, and Sulaiman Ghazi B. Alqahtani

11.1 Introduction

Social networking is now a global trend which has changed the way in which people conduct everyday activities such as social interaction, learning, trading, shopping, leisure time activities and much more. It has been a major influence on people's lives, greatly altering the way people live. Edosomwan, Prakasan, Kouame, Watson and Seymour [18] state that this major change commenced with the emergence of an embryonic form of the web called ARPANET followed by Web 1.0 which came into existence with technologies like WWW, Yahoo! and Google. Technologies underwent further development to produce Web 2.0 which comprises Facebook, YouTube, Twitter and Snapchat, among other applications. With each modified and enhanced iteration of web technology, interactions between individuals and communities have improved in terms of efficiency and effectiveness, enabled by various functions that facilitate the exchanging of ideas and opinions in order to build not only personal communities but also professional groupings.

These technologies are useful not only for leisure activities but also deliver benefits in the business environment, with the help of tools for advertising and marketing processes, thereby improving the way that businesses conduct their commercial operations. Moreover, individuals can keep up to date and engage in a variety of activities from their own homes. These technologies enable people to engage in online shopping, for example. The range of social networking tools has extended to include different sectors like banking, education and health. Moreover, Davis [13] mentions that social networking tools have now been implemented in the

S. Brar · T. Issa (✉) · S. G. B. Alqahtani
Curtin University; School of Management, Perth, WA, Australia
e-mail: Tomayess.Issa@cbs.curtin.edu.au

X. Li, K.-C. Wong (eds.), *Natural Computing for Unsupervised Learning*,
Unsupervised and Semi-Supervised Learning,
https://doi.org/10.1007/978-3-319-98566-4_11

higher education sector because of the teaching and learning benefits that they offer to students, lecturers and tutors.

The availability of social networking tools in higher education delivers various benefits to students and academic staff. Darwish and Lakhtaria [12] point out that universities have incorporated social networking tools in their education setting to improve the delivery of services to students so that they actively interact with their peers and the tutors and can effectively exchange their ideas about any topic through blogs and other tools. The main objective of implementing social networking in the education setting then is to have effective teaching and learning processes. Furthermore, social networking seems to be beneficial for delivering excellent academic outcomes.

However, universities should ensure that they can maximize the benefits to students while minimizing the negative impacts of social networking. This paper examines the risks and benefits of incorporating social networking sites into higher education with reference to India; there is no doubt that there are still some countries which have not effectively incorporated social networking tools in their educational settings because of lack of knowledge [22]. India is one of the countries still struggling with social networking compared to developed countries, although this may no longer be the case as changes occur in culture, society and gender discrimination. This study has collected data from the Indian community and analysed it to determine people's attitudes towards and opinions about social networking especially in the Indian Higher Education. This study will help Indian universities to consider the positive effects of social networking tools in higher education and aiming to minimize the SN negative impacts on the students by adopting specific models and tools especially in the SN implementation in assessments and activities. This study is organized as follows: introduction; social networking, Web 2.0 tools in Indian higher education; SN positive and negative effects; research question; the chosen research method; participants; results; discussion, new findings and recommendations; and conclusion.

11.2 Social Networking

Edosomwan et al. [18] stated that social networks have developed over the years to produce a cutting-edge assortment by utilizing digital media. Online networking technologies are not new. Even more, the techniques have been initiated by the phone rather than a PC. In the 1990s, Six Degrees and Asian Avenue are some interpersonal communication destinations which enabled the individuals to communicate, and it comes under Web 1.0. In 2000 means under Web 2.0, there is an incredible lift with the increase in diverse person to person communication destinations. In 2001, Sky blog and Friendster were introduced. It is demonstrated that the communication innovations are nowadays an important aspect of the education sector, and it is utilized in tertiary education for expanding the connection among undergraduates and teachers and for improving undergraduate skills and individual abilities.

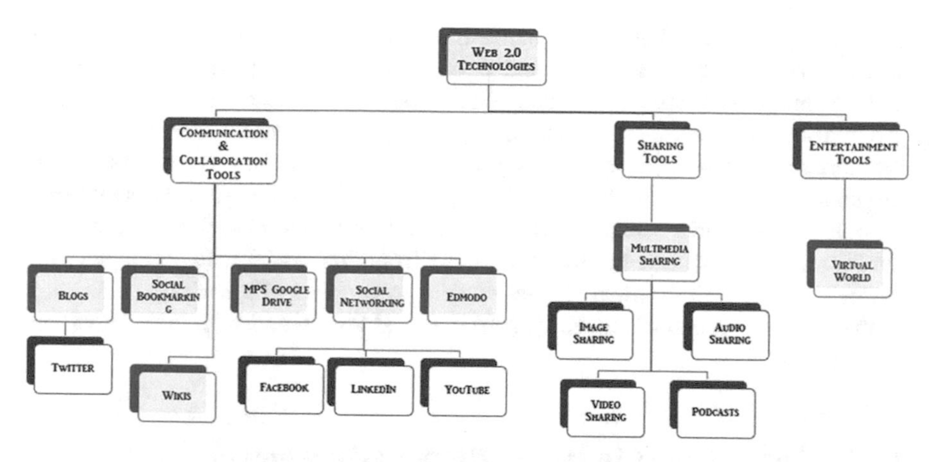

Fig. 11.1 Web 2.0 Typology - prepared by the authors

Binsahl, Chang and Bosua [8] pointed that social networking sites have proven to be very useful for international students for their mobility because these sites are a creative and powerful learning instrument which can be accessed from anywhere at any time. Web-based social networks help students to connect with the whole world easily. Blight, Davis and Olsen [10] stated that the increasing utilization of social media in the education sector by both local and international students has given a new perspective to the way information can be given to students. Web 2.0 is a term that refers to an assortment of sites and operations which enable an individual to make data and then share that data with others. Thomson [43] states that it is the central concept of this innovation to allow individuals to create, share, team up with and impart information or data or material to others so that they can also present their views or further confirm the information.

Figure 11.1 demonstrates the hierarchical nature of the various technology typologies of Web 2.0. Web 2.0 tools are divided into three different categories. First, Ajjan and Hartshorne [1] pointed out that the communication and collaboration tools have the primary motive to enable interaction between people with similar interests. The Wikis, blogs, social networking and social bookmarking are some examples of it. Second, Andersen [3] has explained that multimedia sharing services are the part of the sharing model. It is one of the applications that has seen a great deal of development as it enables people to contribute multimedia content like text, images and videos, and users can comment on others' contributions.

At the end, Marasco, Buonincontri, van Niekerk, Orlowski and Okumus [36] stated that virtual world tool, which is a PC-based group interaction which occurs on the Internet, allows collaboration in a dynamic or custom-manufactured world recreated by people. Web 2.0 applications will play a significant role in changing current teaching and learning practices. Some specific applications which are being used in the higher education sector as displayed in Fig. 11.1 are blogs, wikis, microblogging, media sharing, social networking sites and much more. Blogging is like intelligent diaries where users can, without needing any specialized expertise, create, distribute and organize their pages which contain dated material, sections, contents and exchanges and in chronological order. Moreover, Darwish and Lakhtaria [12] mentioned that Wikis tool is for academic tasks such as group discussions, sorting out recordings and other contributions from group work. Furthermore, microblogging tool is the same as a group of class where students can investigate writings related to group of people, and it is a joint effort crosswise over schools. Besides this, social bookmarking is a tool which allows access to a site via PC which is connected to the web. Apart from this, social networking sites are the most popular tools which enables student to connect with each other. Even, it enables users to remain in touch with friends and family and with their tutors and lecturers.

11.3 Web 2.0 Tools in Indian Higher Education

According to Yadav and A. Patwardhan [50], the utilization of Web 2.0 tools in tertiary education in India is still quite minimal. Considering the high population of India, the extent of implementation is very low compared with some developed countries like Australia, the USA, Canada and others. The number of social network users in India is predicted to double and reach 282 million by 2018. Visits to sites like Facebook, Twitter, YouTube, Pinterest and so forth; communication with friends, relatives, partners and groups; live talk; notifications; images; and video sharing are just some of the activities in which Indians engage online. Indian universities spend approximately 30 min consistently online on specific interpersonal sites. But when it comes to the implementation of Web 2.0 tools in higher education, they are not used to the same extent [23]. IT needs to address a significant number of challenges if universities are to have the high standards of those in developed countries. Universities in India are very aware of the web applications and innovations in software and hardware, and they have a positive attitude towards the utilization of Web 2.0 tools. India is ready to accept web tools.

In Indian universities, the predominant technologies are wikis and social networking tools rather than other tools which are used by students. The reason behind the usage of limited technologies by Indian universities is that not everyone knows about or is familiar with the latest innovations. Tyagi [44] pointed out that some web applications are new even to today's students in India. These tools are also used to provide distance education to students via the Internet. The adaptability and usability are suitable for distance education. However, students and academics are interested in using these technologies in universities for teaching and learning. Yadav and A. Patwardhan [50] stated that there is a need to analyse the capability of these innovations concerning the application of Web 2.0 in e-Learning by considering the underlying achievement factors. They also demonstrated that social networking tools like Facebook give the option to the students for learning the administration of framework.

Tyagi [44] concluded that teachers in India have a negative attitude towards web devices, probably because they have insufficient time for implementation, the

Fig. 11.2 Challenges faced by India in adopting Web 2.0 technologies – prepared by the authors

embarrassment if they don't use technology or have an issue of trustworthiness. Indian students tend to be more receptive to technological innovations than are their teachers. Students are aware of these innovative technologies because they are helping to bring the awareness in academic practices and additionally in observation of instructors and desires.

Yadav and A. Patwardhan [50] mentioned specific difficulties faced by India in adopting the Web 2.0 technology. These are described in Fig. 11.2.

Figure 11.2 depicts some challenges faced by India in adopting Web 2.0 technologies, and they are the limited adoption of ICT, the lack of a reliable infrastructure for Web 2.0 tools, lack of experienced IT personnel, no improvement in the standards and much more.

- There is a limited infiltration of ICTs in semi-urban and rural part of India.
- There is no formal and appropriate course content apart from that in the IT area.
- Students are not able to cope with the lack of a physical presence, which is provided in an actual classroom.
- In India, there is a shortage of teachers and specialists with a sound and comprehensive knowledge of e-Learning and the e-Learning market.
- No practical guidelines are available to the potential user of Web 2.0, and, because of this, its users remain in ignorance of its advantages.

A survey conducted by Verma and Verma [46] found that most respondents were highly familiar with web journals, Google+, Wikipedia, RSS feed, social networks and podcasting. However, apart from the assistant professors, few had any knowledge of SNS, WordPress and Mashups. The researchers also found that the Web 2.0 applications most consistently used by students are Facebook,

Google+, Twitter and social bookmarking, followed by wikis and blogs. Tertiary institutions should be responsible for creating a future information-based society that confidently adopts cutting-edge technology. Tyagi [44] found that gender does not determine the extent to which Web 2.0 technologies are utilized. Web 2.0 is being used in all sectors, including education, and is independent of gender, possibly because of the now ubiquitous nature of web-based social networking.

Tyagi [44] study found that age is a significant determinant of the utilization of Web 2.0, especially in India. The apparent reason is that older Indians have less formal education, whereas the younger generation are becoming educated, are more aware of technologies, and are using them. The majority of Web 2.0 users have higher level qualifications or live in urban areas and are attracted to and aware of Web 2.0 tools. Facebook is the most well-known and popular application in India and worldwide.

The general widespread use of Web 2.0 is in contrast with its utilization in universities; hence, more effort should be made to increase the utilization of Web 2.0 tools for teaching and learning.

Verma and Verma [46] stated that the top institutes of technology in India have adopted the Web 2.0 to some extent. These institutes, together with the technologies that each has adopted, are listed in Table 11.1. This study was conducted on the IIT (Indian Institutes of Technology) institutes which are the most popular and top institutes in India for delivering courses in technology. Table 11.1 shows the utilization of a variety of Web 2.0 tools by the top institutes of technology in India. Table 11.1 shows the utilization of Web 2.0 in IITs. Currently, only ten IITs are utilizing Web 2.0 tools, and five IITs (Kharagpur, Guwahati, Ropar, Patna and BHU) are not aware of the utilization of any Web 2.0 instrument. The greatest number of technologies is used by the IIT Bombay, followed by the IIT Hyderabad and the IIT Roorkee. IIT Gandhinagar and IIT Madras that are using four and, furthermore, three Web 2.0 tools are being used by IIT Indore. Only two Web 2.0 technologies are currently being used by IIT Delhi and IIT Jodhpur. But only one Web 2.0 tool is being used by IIT Mandi.

Verma and Verma [46] found that the IITs of Varanasi, Patna, Ropar, Guwahati and Kharagpur have not adopted a single Web 2.0 technology for academic activities; hence, they may not feel comfortable using the latest technologies. However, this research study does not indicate the number of the followers or likes for every specific tool. The most popular tools used in IIT institutes are Google+, Facebook, Twitter and YouTube and the least utilized tools are Wikipedia and WordPress. The researchers also found that students and tutors were comfortable with exchanging their academic material or views regarding any field via social networking tools like Facebook and Twitter because of their flexible and easy accessibility rather than using WordPress and Wikipedia. However, despite the findings of this particular study, there is still no clear reason why some institutions have not adopted Web 2.0 technologies or how this situation can be improved to meet the current technology needs of the education sector. It is also unclear whether the current level of utilization was adequate for those institutes that had adopted some Web 2.0 tools.

Table 11.1 Web 2.0 tools usage in Indian higher education – prepared by the authors

Name of the institute	Facebook	Twitter	Google+	LinkedIn	YouTube	Blog	Wikipedia	WordPress	RSS	Total
IIT, Kharagpur	–	–	–	–	–	–	–	–	–	0
IIT,Bombay	Yes	Yes	Yes	Yes	Yes	–	–	–	Yes	6
IIT, Madras	Yes	Yes	Yes	Yes	–	–	–	–	–	4
IIT Delhi	–	–	–	–	Yes	–	–	–	Yes	2
IIT, Guwahati	–	–	–	–	–	–	–	–	–	0
IIT, Roorkee	Yes	Yes	Yes	Yes	–	Yes	–	–	–	5
IIT, Hyderabad	–	Yes	–	–	Yes	Yes	–	Yes	Yes	5
IIT, Jodhpur	–	–	Yes	–	–	Yes	–	–	–	2
IIT, Ropar	–	–	–	–	–	–	–	–	–	0
IIT, Mandi	Yes	–	–	–	–	–	–	–	–	1
IIT, Indore	–	–	Yes	–	–	–	Yes	–	Yes	3
IIT, Patna	–	–	–	–	–	–	–	–	–	0
IIT, Gandhinagar	Yes	Yes	Yes	–	Yes	–	–	–	–	4
IIT,Bhubaneswar	–	–	–	–	Yes	–	–	–	–	1
IIT,Varanasi	–	–	–	–	–	–	–	–	–	0
Total	5	5	6	3	5	3	1	1	4	

11.4 SN Positive and Negative effects

SN tools have some positive and negative. The positive of these tools are helpful to lower costs, no need to have a high level of expertise in order to use these technologies, access data higher speed, adaptable in nature and much more. Some negative of these are Internet connection required, dependent on Ajax and JavaScript may contain unreliable or incorrect content, face some security concerns while conveying sensitive information and time and resources are required. These tools also have some positive and negative effects which are discussed in detail in the following and displayed in Fig. 11.3. The positive effects are that they offer cutting-edge knowledge which means Web 2.0 tools like a social network can deliver information which is not necessarily restricted to one area of interest or issue. Al-Rahmi and Zeki [2] pointed that social networking can help students to develop collaboration skills. It provides an avenue for students to efficiently get in touch with each other about class tasks, group assignments or homework by means of sites like Facebook, Twitter and others. Social networking can improve students' learning outcomes and encourages cooperative learning in tertiary institutions. Investment in social networking can support learning. Therefore, as the interest in dynamic community learning expanded, researchers and analysts began to adopt online networking. With the help of social networking, students become more confident about discussing issues with others and addressing issues in a community.

Social media helps students to make friends and allows them to communicate with one another and share their ideas on any topic. It can improve students'

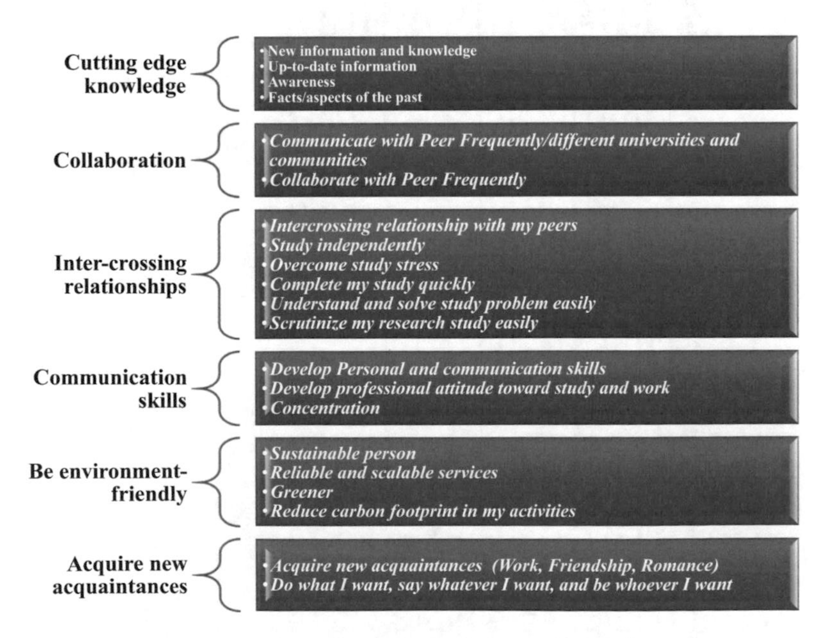

Fig. 11.3 Positive effects of social networking – prepared by authors

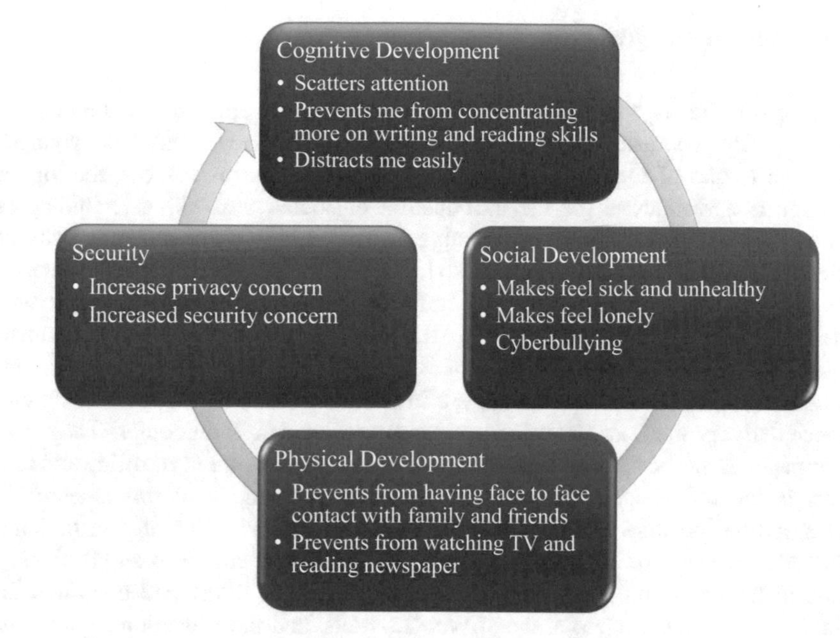

Fig. 11.4 Negative effects of social networking – prepared by authors

dependability as group members come to rely on one another. Furthermore, opportunity to make new acquaintances means online networking is something that has permeated all aspects of our lives, from family and connections through to shopping and world news and issues.

Some negative effects of social networking tools as per Osborne [40] are on cognitive development, which means lack of concentration on the studies by students because they are more comfortable communicating via social media, which is a distraction. Furthermore, it affects the social development by having excessive use of these tools which further lead to face mental and physical health issues.

Figure 11.4 illustrates some risks of using social networking sites and showed different parameters related to each negative aspect. Moreover, it increased the privacy and security concerns by enabling individual to gain unauthorized access to a website. Besides this, Lederer [32] pointed that the physical development of student is affected because some students are too shy to communicate directly with others and come to rely on social networks for their interaction with others.

The journal has provided the comprehensive review on the use of social networking utilization in higher education and examined the positive and negative effects of social networking tools in higher education by investigating the significant patterns which have emerged over the previous decade. It concluded that the utilization of social networking tools has positive as well as negative side, but these depend on the ways in which these tools are used and to what extent.

Finally, Figs. 11.3 and 11.4 demonstrated SN positive and negative based on the current literature [4–7, 14–17, 19, 21, 24, 26, 28, 30–35, 37–39, 41, 42, 45, 47, 48].

11.5 Research Question

Social networking is without a doubt the most common approach to facilitating communication, connection and collaboration with others. In particular, the younger generation is one of the top users of social networks. In this respect, the higher education is a ripe arena for the introduction of social networking technologies for the purposes of teaching and learning. However, there are inherent risks in implementing social networking sites into higher education, particularly in societies such as India where conservative and traditional values prevail. This paper will explore and therefore pose the following question: what are the positive and negative of using social networking sites in India's higher education sector? In so doing, it examines the current education system in India and how social networking can enhance delivery methods, learning processes and academic outcomes. The focus of the paper is on how these benefits can be harvested while controlling for risk factors in the university environment. The primary objective of this research is to examine the positive and negative of using social networking sites in Indian's higher education sector. India is still one of the developing countries and is lagging behind in financial and social development compared to developed countries. In India, there is an insufficient usage of Web 2.0 tools, and the education department is still immersed in a traditional way of delivering education. However, education authorities are trying to incorporate social networking tools in order to provide an educated and skilled workforce to the industries of India. Based on this, India has set out to encourage its education department to take an interest and embark on change. Already some noteworthy changes have been made to achieve the goal of providing quality education that meets international standards.

11.6 The Chosen Research Method

The quantitative research method was chosen to collect and analyse the data for this paper. The online survey was developed based on the current literature and divided into three sections, namely, background, positive and negative sections. In the positive section, there were 25 statements targeting statements related to cutting-edge knowledge, collaboration, intercrossing relationships, communication skills, be environment-friendly and acquire new acquaintances. On the other hand, in the negative section, there were 30 statements aiming for cognitive development, social development, physical development and security. Figures 11.3 and 11.4 are presenting the online survey questions. The online survey was distributed via the Qualtrics platform to the Indian students in higher education to examine SN awareness, and the duration for online survey data collection was 2 months. Finally, five-point Likert scale (1 = not at all to 5 = extremely) was usedfor the positive and negative statements in the survey to assess the intensity of a respondent's perceived approval or disapproval towards the statements [9, 25].

It was intended to deliver quantifiable and numerical information to gauge the positive and negative impacts of social networking in higher education in India. The data was collected by means of a survey administered to numerous individuals residing in India. After collection, the data was subjected to the expository technique which reveals common factors in the data and enables the comparison of mean values. The online survey was conducted to gather the primary data. This is a convenient way of gathering valid data about attitudes and opinions, behaviours and situations. The online survey was conducted to collect the attitudes of respondents regarding the positive and negative effects of social networking in India. The online survey was distributed by various means throughout the Indian community, and responses were collected and then analysed with the help of systematic quantitative methods.

The web-based survey can be used to gather data from respondents by distributing the questionnaires via Facebook, WhatsApp, Twitter and much more. Issa [29] states that the utilization of the web-based survey has several advantages over traditional survey methods such as the pen or paper-based survey or email. The online survey is easy to use, of low cost and efficient to distribute, and responses can be submitted online by anyone. Furthermore, respondents can be regularly reminded to complete the survey. Moreover, the survey can be opened or downloaded in one's preferred format. The flexible design means that response rates can be monitored, and the data can be analysed more easily than with conventional methods.

However, Issa [29] claims that the online survey does not only provide advantages but also has some disadvantages, one of which is that respondents might not understand the survey instructions correctly. Notably, this occurs in the case of a sophisticated survey design. This disadvantage was experienced when carrying out the survey. Because the data was sought from the Indian community, it was difficult to obtain an adequate number of valid responses since Indians are still lagging behind when it comes to understanding technology and being able to access social networking tools. Thus, it was difficult to collect responses from a total of just 100 respondents.

The data analysis was conducted using the Social Science Statistics Software (SPSS) 23.0, and Qualtrics was used for the survey. Zueger, Katz and Popovich [51] suggest that Qualtrics is an easy-to-use electronic survey tool to direct overall research, assessments and other information-gathering exercises. The use of this tool requires no experience, and an individual can design and distribute surveys and monitor the responses. Furthermore, another statistical tool used for this paper for data analysis is descriptive statistics and factor analysis. These tools can help to provide new understanding in respect of the risks and advantages of utilizing social networking as perceived by Indian people. It is also an effective means of discovering the benefits which social networking tools can deliver to the higher education sector in India.

11.7 Participants

Table 11.2 shows the percentages of male and female survey respondents. Out of 142 participants, 51.4% of male participants and 48.6% of female participants returned the data for the survey. Table 11.2 shows the percentage of participants utilizing the Internet based on their age group. The people in the 22–32-year range are the highest users of the Internet. The least number of users is in the over 52 age group and those between 42 and 52. Moreover, 14.1% and 14.8% of participants had submitted their responses, and they were in the 32–42 and 18–22 age groups, respectively.

The highest number of participants (16.2%) has careers in the field of management, followed by those in accounting and marketing (13.4%). Those in the computer science and information technology fields account for 12.7% and 11.3%, respectively. Table 11.1 shows that for the majority of participants (36.6%), the highest level of education reached was the diploma qualification. The second highest level was the bachelor's degree which was held by 34.5% of the participants. 14.1% of the participants hold master's degrees. Moreover, the lowest percentages of the participants had completed a postgraduate diploma, higher secondary and primary education, advanced diploma and professional certificate at 1.4%, 2.1%, 3.5% and 2.8%, respectively.

It can be seen from Table 11.2 that the majority (55.6%) of participants spent up to 5 h on social networking sites except for email. 20.4% of the participants like to utilize social networking sites up to 5–10 h daily. 17.6% spent less than an hour on social networking sites. The lowest numbers of respondents were those who spent 10–20 h and over 20 h on social networking which accounted for 3.5% and 0.7% of respondents, respectively.

Table 11.1 shows the percentage of hours spent by participants on email each day except for social networking tools. Most of the participants spend time less than an hour and up to 5 h in a day on email which accounted for 42.3% and 40.8%, respectively. Only 12.7% of participants spend 5–10 h on email. The least numbers of participants (0.7%) are those who spend 10–20 h and over 20 h on email each day.

11.8 Results

A total of 25 items were assessed for consistency in terms of the positive effects of social networking from Indian people's perspective. The results above indicate that the model which consists of positive characteristics is exceptionally predictable (consistent) because Cronbach's Alpha is 0.919. Gliem and Gliem [20] mentioned in their study that if Cronbach's Alpha is greater than 0.9, then outcomes are consistent. There were 30 segments which were assessed to establish the reliability of the negative aspects of social networking which emerged from the survey. From the above, one can see that the negative effects were highly predictable as the model for the negative effects had a Cronbach's alpha of 0.914 [20] (see Table 11.3).

Table 11.2 Online survey statistics India – prepared by the authors

	Frequency	Percent
Gender		
Male	73	51.4
Female	69	48.6
Age		
18–22	21	14.8
22–32	100	70.4
32–42	20	14.1
42–52	1	0.7
Field(s) of study		
Accounting	19	13.4
Business law	4	2.8
Economics and finance	6	4.2
Information systems	5	3.5
Information technology	16	11.3
Computer science	18	12.7
Management	23	16.2
Marketing	19	13.4
Health sciences	4	2.8
Humanities	2	1.4
Science and engineering	10	7.0
Art and design	9	6.3
Others – please specify	4	2.8
Field(s) of education level		
Primary education	5	3.5
Higher secondary/preuniversity	3	2.1
Professional certificate	5	3.5
Diploma	52	36.6
Advanced/higher/graduate diploma	4	2.8
Bachelor's degree	49	34.5
Postgraduate diploma	2	1.4
Master's degree	20	14.1
Hours you spend on the social networking daily, not including email (per day)		
Less than an hour	25	17.6
Up to 5 h	79	55.6
5–10 h	29	20.4
10–20 h	5	3.5
Over 20 h	1	0.7
Hours you spend on the Internet for email (per day)		
Less than an hour	60	42.3
Up to 5 h	58	40.8
5– 10 h	18	12.7
10–20 h	1	0.7
Over 20 h	1	0.7

Table 11.3 SN statistics (alpha, KMO and Bartlett's test) – prepared by the authors

Factor	Cronbach's alpha	KMO sampling adequacy	Bartlett's test of sphericity
Positive	0.919 [excellent]	0.862 (meritorious)	$\chi\hat{}2 = 1029.526$; df = *300* $p < 0.000$
Negative	0.914 [excellent]	. 821 (meritorious)	$\chi\hat{}2 = 1141.809$; df = *435* $p < 0.000$

Table 11.4 SN positive effects – descriptive statistics – prepared by the authors

	Mean	Std. deviation
Learn new information and knowledge	3.97	1.050
Acquire new acquaintances – work related	3.74	0.988
Acquire new acquaintances – friendship relationship	3.68	0.904
Communicate with my peers frequently	3.69	1.009
Provide reliable and scalable services	3.69	1.019
Complete my study more quickly	3.67	0.972
Concentrate more on my reading and writing skills	3.66	1.064
Communicate with my different communities	3.64	0.966
Study independently	3.58	0.994
Understand and solve study problems easily	3.57	0.995
Scrutinize my research study more easily	3.62	1.000
Develop my personal and communication skills	3.61	0.938
To prepare my professional attitude towards study and work	3.59	1.054
Be more sustainable person	3.55	0.996
Become "greener" in my activities	3.53	1.133
Reduce carbon footprint in my activities	3.51	0.933
Acquire new acquaintances – romance relationship	3.50	1.096
Do whatever I want, say whatever I want and be whoever I want	3.50	1.038
Develop intercrossing relationships with my peers (i.e. artistic talents, sport and common interests)	3.46	0.965
Overcome study stress	3.49	0.944

The Kaiser-Meyer-Olkin (KMO) measure of sampling adequacy value is 0.862 which is higher than 0.7, and it stands for excellent. The Bartlett's test of sphericity value will be significant if it is less than (0.05). The above results demonstrate that the KMO value is (0.862) and the Sig. value is (0.000). Therefore, factor analysis is suitable for this case as explained by Williams, Onsman and Brown [49] in their work on exploratory factor analysis (see Table 11.2). The Kaiser-Meyer-Olkin (KMO) measure of sampling adequacy [20] value is 0.821 which is higher than 0.7, and it stands for excellent. It can be accurate to say that the sampling adequacy for the test is met. The Bartlett's test of sphericity value will be significant if it is less than 0.05. The above results demonstrate that the KMO value is 0.862 and the Sig. value is 0.000. Therefore, factor analysis is appropriate for this case as mentioned by Williams et al. [49] in their work on exploratory factor analysis (see Table 11.2).

Table 11.4 presents the results for the positive effects of social networking. The first six factors have the highest mean values compared to the other elements. The primary factor which is "Learn new information and knowledge" has the highest mean value of 3.97 with 1.050 standard deviation. The lowest mean value was for the factor "Develop intercrossing relationships with my peers (i.e. artistic talents, sport and common interests)" which was 3.46 in comparison with all other factors with a standard deviation of 0.965.

Table 11.5 shows the descriptive statistics for the negative effects. The first six factors have the highest mean values compared to the other factors. The primary factor which is "Prevents me from reading the newspapers" has the highest mean value which is 3.58 with a standard deviation of 1.004. The lowest mean value is for the factor "Depresses me" which was 3.11 in comparison with all factors, with a standard deviation of 0.994.

Table 11.5 SN negative effects – descriptive statistics – prepared by the authors

	Mean	Std. deviation
Prevents me from reading the newspapers	3.58	1.004
Prevents me from talking on the phone/mobile	3.49	0.876
Distracts me easily	3.47	1.017
Increase security concerns	3.45	0.875
Prevents me from completing my work/study on time	3.40	0.939
Scatters my attention	3.24	0.953
Decreases my grammar and proofreading skills	3.24	0.985
Decreases my deep thinking	3.23	1.053
Distracts me easily	3.47	1.017
Stresses me	3.33	1.091
Makes me feel lonely	3.26	1.106
Makes me lazy	3.36	1.133
Makes me addict	3.31	0.968
Makes me insecure to release my personal details from the theft of personal information	3.32	1.041
Makes me receive an immoral images and information from unscrupulous people and it is difficult to act against them at present	3.26	0.912
Prevents me from having face-to-face contact with my family	3.30	0.922
Prevents me from having face-to-face contact with my friends	3.37	0.901
Prevents me from participating in physical activities	3.36	0.977
Prevents me from shopping in stores	3.38	0.990
Prevents me from watching television	3.39	0.857
Prevents me from completing my work on time	3.28	0.972
Prevents me from completing my study on time	3.33	0.928
Increase privacy concerns	3.36	0.987
Increase intellectual property concerns	3.31	0.842
Prevents me from participating in social activities	3.21	1.008
Bores me	3.24	1.046
Makes me sick and unhealthy	3.18	0.988
Makes me more gambler	3.13	1.061
Depresses me	3.11	0.994

Blunch [11] has defined factor analysis as an attempt to recognize the fundamental factors which demonstrate a pattern of connections in between the pattern of observed factors for the positive and negative effects of social networking. Factor analysis is regularly utilized to condense the amount of information into a few significant factors which clarify the more primary part of the fluctuation. It permits a great number of factors to be effortlessly condensed to a smaller number which can be more easily tested, and these can then be utilized as a part of different tests.

Williams et al. [49] stated that while doing the extraction, the data is considered in depth for deciding the number of elements. The following results are for the total variance explained, and it was utilized for the components which had an eigenvalue of 1.0 or more, and the values which are lower than 1.0 are excluded. In the following results, the outcomes which met this measure are demonstrated. The first three columns show the values for 25 factors. The factors from 1 and 2 have values greater than 1.0 and this constitutes 41.280% of the total variance (see Table 11.6).

Table 11.7 is the rotated component matrix for the positive factors. The values contained in it are the two factors which have been mentioned previously in the total variance explained. The components of the different factors can be categorized and highlighted in order to generate a new factor from a group of factors. This is done for each component factor.

From the rotated component matrix shown in Table 11.6, the components which comprise different factors are grouped together, and positive new factors are generated based on the Indian higher education, namely, developing personal skills and communicate and accessible service (see Table 11.8).

Table 11.9 discusses the total variance explained for the SN negative factors; it was utilized to retain the components which had an eigenvalue of 1.0 or more, and the values which are lower than 1.0 are discarded [50]. The following results show the outcomes of this measure. The first three columns contain the values for 30 factors. The factors from 1 to 5 have values greater than 1.0, comprising 51.052% of the total.

The results in Table 11.10 are for the rotated component matrix – negative factors. The values are for the five factors which were also mentioned previously in the total variance explained section. The components of the different factors can be categorized and highlighted for the purpose of generating a new factor. This is done for each component.

From the rotated component matrix as shown in Table 11.10, the components of the different factors are grouped together, and a new factor is generated from them. Moreover, when analysing the data, it was noticed that some of the participants had provided additional comments. The newly generated factors for the negative are shown in Table 11.11.

Table 11.6 Total variance explained – SN positive factors – prepared by the authors

Component	Initial eigenvalues			Extraction sums of squared loadings			Rotation sums of squared loadings		
	Total	% of variance	Cumulative %	Total	% of variance	Cumulative %	Total	% of variance	Cumulative %
1	8.613	34.454	34.454	8.613	34.454	34.454	5.165	20.660	20.660
2	1.707	6.826	41.280	1.707	6.826	41.280	5.155	20.620	41.280

Extraction method: principal component analysis

Table 11.7 Rotated component matrix – SN positive factors – prepared by the authors

Rotated Component Matrix		
	Component	
	1	2
Develop my personal and communication skills	.686	
Collaborate with my peers frequently	.639	
Study independently	.635	
Communicate with my different communities	.630	
Acquire new acquaintances - romance relationship	.600	
Communicate with my peers from different universities		.752
Provide reliable and scalable services		.680
Scrutinize my research study more easily		.667
Be more aware of global issues/local issues		.662
Overcome study stress		.618

Extraction method: principal component analysis
Rotation method: varimax with Kaiser normalization
a. Rotation converged in 3 iterations

Table 11.8 New SN positive factors – India – prepared by the authors

Variable Name	**Factor loading**	**New factor**
Develop my personal and communication skills	.686	Developing personal skills
Collaborate with my peers frequently	.639	
Study Independently	.635	
Communicate with my different communities	.630	
Acquire new acquaintances- romance relationship	.600	
Variable Name	**Factor loading**	**New factor**
Communicate with my peers from different universities	.752	Communicate and accessible service
Provide reliable and scalable services	.680	
Scrutinize my research study more easily	.667	
Be more aware of global issues/local issues	.662	
Overcome study stress	.618	

11.9 Discussion, New Findings and Recommendations

The developments and innovations of web tools have brought many benefits to the lives of individuals and communities, and improvements are constantly being made to technology. The web-based technologies evolved from Web 0.0 to Web 2.0 tools. Each technology delivers effective results to their end-users, but advancement from Web 1.0 to Web 2.0 has facilitated quicker and more effective correspondence between individuals. The human interaction is not limited to one sector; with Web 2.0 tools, communication can occur in multiple areas like trade, banking, business, education and recreational activities. Social networking tools are useful for entertaining individuals and alleviating boredom.

The primary objective of this study is to discover whether the Web 2.0 tools which offer multiple features, facilitate effective communication and enhance knowledge and much more are effective in the higher education sector. These tools can be utilized for multiple purposes like dynamic learning development and organizing study groups online. Moreover, valuable information can be exchanged, and students can more easily interact with their peers and teachers, by means of the

Table 11.9 Total variance explained – negative – prepared by the authors

Total variance explained									
Component	Initial eigenvalues			Extraction sums of squared loadings			Rotation sums of squared loadings		
	Total	% of variance	Cumulative %	Total	% of variance	Cumulative %	Total	% of variance	Cumulative %
1	8.745	29.151	29.151	8.745	29.151	29.151	3.432	11.440	11.440
2	1.944	6.480	35.632	1.944	6.480	35.632	3.251	10.837	22.277
3	1.767	5.889	41.520	1.767	5.889	41.520	3.108	10.360	32.637
4	1.505	5.017	46.537	1.505	5.017	46.537	2.964	9.881	42.519
5	1.355	4.515	51.052	1.355	4.515	51.052	2.560	8.534	51.052

Extraction method: principal component analysis

Table 11.10 Rotated component matrix – SN negative – prepared by the authors

	Component				
	1	2	3	4	5
Makes me more gambler	.682				
Makes me feel lonely	.668				
Bores me	.605				
Makes me sick and unhealthy	.596				
Makes me addict	.551				
Stresses me	.501				
Prevents me from reading the newspapers		.635			
Prevents me from participating in physical activities		.587			
Scatters my attention		.553			
Prevents me from having face to face contact with my family		.523			
Prevents me from completing my work/study on time		.518			
Prevents me from participating in social activities		.511			
Makes me lazy			.725		
Depresses me			.573		
Makes me receive an immoral images and information from unscrupulous people and it is difficult to act against them at present			.565		
Makes me insecure to release my personal details from the theft of personal information			.555		
Distracts me easily			.540		
Prevents me from talking on the phone/mobile				.687	
Increase intellectual property concerns				.643	
Prevents me from completing my work on time				.583	.
Prevents me from having face to face contact with my friends				.523	
Prevents me from shopping in stores				.519	
Decreases my grammar and proofreading skills					.740
Prevents me from remembering the fundamental knowledge and skills					.696
Prevents me from concentrating more on writing and reading skills					.599
Increase security concerns					.550

Extraction method: principal component analysis
Rotation method: varimax with Kaiser normalization
a. Rotation converged in 24 iterations

various available online tools. For the universities, they are not limited to the extent that students interact with these tools and utilize them in their study work. Those responsible for delivering education should also be concerned with incorporating these tools in teaching and learning activities in order to improve student outcomes and make them successful in an increasingly competitive world.

The whole study focused on this mentioned objective and gained in-depth insights into the positive and negative effects of social networking tools in higher education so that universities will consider the advantages and opportunities offered by Web 2.0 technologies rather than considering only the dangers and risks. For this study, the online survey was distributed to Indian people, and all valid responses were analysed and evaluated. The Indian community has presented their perception on the positive and negative effects of social networking tools in higher education. This study reveals the current situation regarding the implementation of social networking tools in higher education and how this could improve and increase the utilization of Web 2.0 tools because India is still far behind in the development and usage of tools in higher education. Figure 11.5 shows the positive factors and sub-variables related to the social networking effects that emerged from the data analysis. These factors are described in more detail below.

Table 11.11 New SN negative factors – India – prepared by the authors

Variable Name	Factor Loading	New Factor
Makes me more gambler	.682	Negative emotions created
Makes me feel lonely	.668	
Bores me	.605	
Makes me sick and unhealthy	.596	
Makes me addict	.551	
Stresses me	.501	
Variable Name	**Factor Loading**	**New Factor**
Prevents me from reading the newspapers	.635	Decrease productivity in social activities
Prevents me from participating in physical activities	.587	
Scatters my attention	.553	
Prevents me from having face to face contact with my family	.523	
Prevents me from completing my work/study on time	.518	
Prevents me from participating in social activities	.511	
Variable Name	**Factor Loading**	**New Factor**
Makes me lazy	.725	Laziness and Depression issues
Depresses me	.573	
Makes me receive an immoral images and information from unscrupulous people and it is difficult to act against them at present	.565	
Makes me insecure to release my personal details from the theft of personal information	.555	
Distracts me easily	.540	
Variable Name	**Factor Loading**	**New Factor**
Prevents me from talking on the phone/mobile	.687	Lack of Communication and IP
Increase intellectual property concerns	.643	
Prevents me from completing my work on time	.583	
Prevents me from having face to face contact with my friends	.523	
Prevents me from shopping in stores	.519	
Variable Name	**Factor Loading**	**New Factor**
Decreases my grammar and proofreading skills	.740	Decrease in study skills
Prevents me from remembering the fundamental knowledge and skills	.696	
Prevents me from concentrating more on writing and reading skills	.599	
Increase security concerns	.550	

Developing personal skills: The factors under this category are "Acquire new acquaintances – romance relationships", "Developing my personal skills", "Collaborate with peers frequently", "Study independently" and "Communicate with my different communities", each of which had a factor loading of 0.6.

Communicate and accessible services: This comprises five factors of which "Communicate with peers from different universities" has the highest factor loading of 0.7, and other remaining factors have the same factor loading which is 0.6. The remaining factors are "Provide reliable and scalable services", "Be more aware of global issues/local issues", "Overcome study stress" and "Scrutinize my study more quickly".

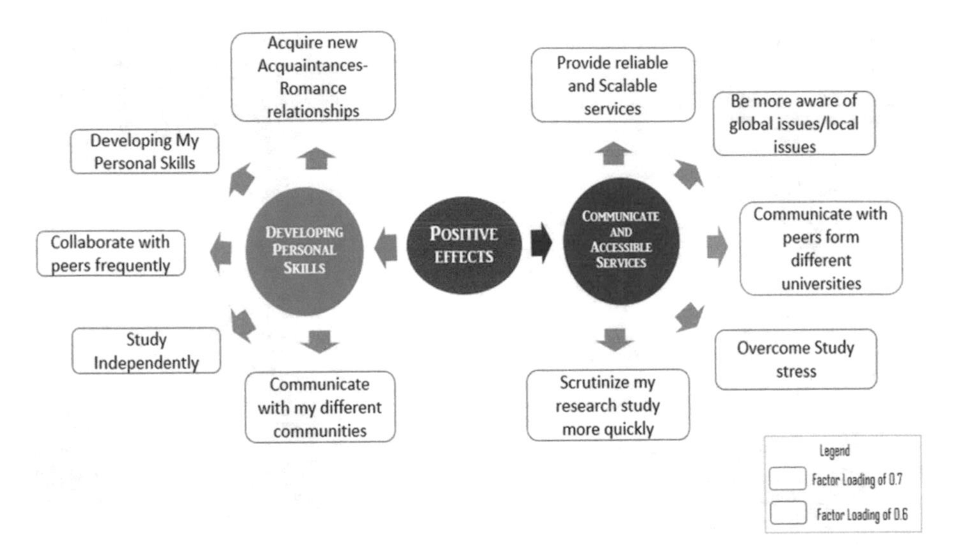

Fig. 11.5 Positive factors of social networking – prepared by the authors

Figure 11.5 illustrates the factors which emerged for the positive effects of social networking. These factors were generated by an analysis of the survey data. The factor with the highest loading of 0.7 was "Communicate with peers from different universities". Ajjan and Hartshorne [1] stated that social networking tools deliver various benefits to individuals such as communication, collaboration, development of skills and a more effective means of study. The factor with the highest factor loading for the positive effects of social networking tools is "Communicate with peers form different universities". This indicates that, in India, social networking tools are highly useful for communication related to study; this may be because of the restricted environment in India which means that students cannot effectively communicate face-to-face with peers, so they prefer social networking tools for it.

Moreover, other factors like "Overcome study stress", "Scrutinize my research study more quickly" and "study independently" indicate that social networking tools are highly useful in the education field in India, compared to other factors. It was noted that it is an effective means of making people aware of global/local events which keeps people up to date. Apart from this, the findings indicated that the Indian community uses social networking tools to make new acquaintances or to establish a romantic relationship, which is entirely different from other factors. This factor may be explained by the fact that India has a strict social code that prevents young people especially from communicating face-to-face, so they prefer to use social networking tools to become acquainted.

It can be concluded that social networking tools are being extensively used for educational purposes. Hence, Indian universities should be encouraged to adopt these tools in their curriculum, confident that they will deliver expected, or better, performance outcomes.

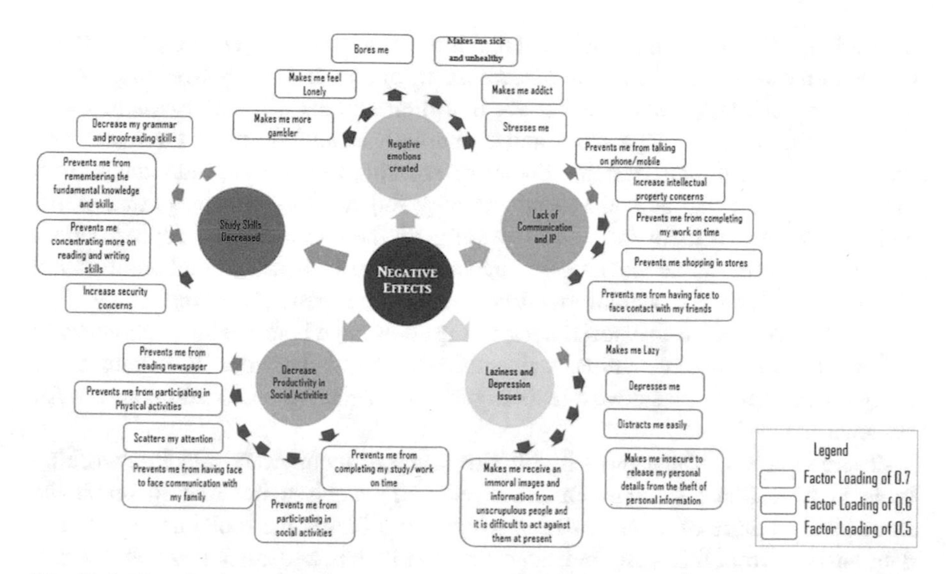

Fig. 11.6 Negative factors of social networking – prepared by Shamdeep Brar

Figure 11.6 depicts the negative factors which emerged from the analysis of the survey data. It also includes the new factors that were generated. The factors and their loadings are described below.

Negative emotions created: Here, three factors had a factor loading of 0.6: "Bores me", "Makes me feel lonely" and "Makes me gambler"; the other factors had a factor loading 0.5.

Study skills decreased: The factor with the highest factor loading of 0.7 was "Decrease my grammar and proofreading skills". The two factors "Prevents me from remembering the fundamental knowledge and skills" and "Prevents me concentrating more on reading and writing skills" received a factor loading of 0.6. The lowest factor loading in this category was 0.5 for the "Increase security concerns" factor.

Decreased productivity in social activities: The highest factor loading under this category was 0.6 for "Prevents me from reading the newspaper". The remaining factors had the same factor loading of 0.5.

Laziness and depression issues: This category had the highest factor loading, i.e. 0.7, for the "Makes me lazy" factor, and other remaining factors had the same lowest factor loading of 0.5.

Communication issues: The factors of "Prevents me from talking on phone" and "Increase intellectual property concerns" had the same factor loading of 0.6, and the other remaining factors under this category had the same factor loading of 0.5.

Figure 11.6 depicts the negative factors which emerged from the analysis of the data related to Indian people's attitudes towards and perceptions of social networking. The survey data indicated that individuals perceive two negative effects of social networking; these were "Decrease in grammar and proofreading skills" and "Makes me lazy". Students believe that their studies will suffer as a result of a decrease in their language skills, and social networking will make them lazy and distract them from their learning. Social networking can prevent individuals from participating in social activities; students may forget basic knowledge and skills, and it may prevent them from concentrating on their studies. The concern about security is another issue that emerged because individuals do not feel comfortable about sharing their personal information when using social networking tools.

It can be concluded that social networking tools have both positive and negative effects. Before implementation, universities should take steps to eliminate or mitigate the potential negative effects in order to achieve the desired outcomes for students and staff.

There is no doubt that India is still struggling to incorporate social networking tools in educational settings. Far more effort is required in order to reach the same level as those of other countries which are utilizing the social networks. The data analysis from this study indicated that the Indian students know about social networking and know that it can benefit their education; however, they are unware of SN negative effects; therefore, to minimize these effects, the Indian universities should implement SN tools in their curriculum, especially in the assessments and activities by adopting SN model, i.e. Social Networking and Education Model (SNEM) aims to promote and implement Social Networking in the education sector to increase positive SN effects and reduce negative SN effects [27]. This study confirmed the study aims and research question and added new significance and findings to the current literature, especially to SN awareness and implementation especially in India.

This section comprises several recommendations for a better understanding of the positive and negative effects of social networking; they are explained as follows:

- The positive and negative effects of social networking need to be known so that the universities will have a clear idea about how social networking can be effectively implemented. Basically, this needs to be done for motivating or encouraging the universities to implement them as well as enrich the knowledge relevant to the utilization of social networking tools in higher education.
- Indian universities lack effective Internet access and web platforms for their students as they do not provide online accounts enabling student access to various tools. So, it is recommended that for effective implementation of social networking tools in universities, they need to provide Internet access without which these tools cannot be used. Moreover, the assignment of online accounts will help the students to access the tools free of cost which will enhance the reputation of the university.
- It is recommended that universities organize seminars for students and staff to make them aware of the utilization of social networking tools, especially since older staff may be unaware of networking tools and how to use them. Moreover, training is another recommendation for making people familiar with the tools

which can be further utilized in the education curriculum. This will help the individual to build confidence in utilizing the tools effectively, and the rate of user nonacceptance will decrease.

- After the evaluation of negative effects of social networking tools, it was noticed that individuals are feeling insecure about sharing their personal information on the Internet. So, it is recommended that effective measures be implemented to safeguard individuals' personal information from unauthorized access.
- In India, social networking tools have been adopted mostly by the top universities, and some universities are still unaware of them. So, it is highly recommended that steps be taken to spread the awareness of social networking tools and how these tools can benefit students' education by making them aware of local/global issues, and making them more sustainable, and much more. It will help to enrich the Indian community's knowledge of social networking tools utilization.

Numerous research restrictions are recognized and contain challenges related to time, sample size and difficulties in contacting potential respondents. In the future to confirm the study aims and objectives, additional research will be carried out to examine the study aims with more varied cluster of participants in India.

11.10 Conclusion

This study aims to study the effects of social networking in higher education in India; the online survey results generated new positive factors, namely, developing personal skills and communicate and accessible service, while the negative factors were, namely, negative emotions created, decreased productivity in social activities, laziness and depression issues, lack of communication and IP and decrease in study skills. The positive effects have considered that Indian students use SN tools to develop their personal skills and improve their communication locally and globally. However, this tool created several negative effects which are considered a warning message and action call to the Indian higher education to implement social networking tools in assessments and activities to motivate and encourage students to learn the necessary skills from these tools which are required for their study and workforce in the future and the more important to reduce the negative effects of SN. Therefore, a social networking model should be implemented to minimize the negative and increase the positive effects in higher education in India. Finally, further research will be carried out to examine the study aims with more diverse groups in India.

References

1. Ajjan H, Hartshorne R (2008) Investigating faculty decisions to adopt Web 2.0 technologies: theory and empirical tests. Internet High Educ 11(2):71–80
2. Al-Rahmi WM, Zeki AM (2016) A model of using social media for collaborative learning to enhance learners' performance on learning. J King Saud Univ Comput Inf Sci 29(4): 526–535
3. Andersen P (2007) What is Web 2.0?: ideas, technologies and implications for education. JISC Bristol 1:1
4. Aral S (2013) Social media and business transformation: a framework for research. Inf Syst Res 24(1):3–13
5. Arlow P (1991) Personal Characteristics in College Students' Evaluations of Business Ethics and Corporate Social Responsibility. J Bus Ethics 10(1):63–69
6. Baams L, Jonas K, Utz S, Bos H, Vuurst L (2011) Internet use and online social support among same sex attracted individuals of different ages. Comput Hum Behav 27(5):1–8
7. Bandura A (1977) Social learning theory. Holt, Rinehart and Winston, New York
8. Binsahl H, Chang S, Bosua R (2015) Identity and belonging: Saudi female international students and their use of social networking sites. Cross J Migr Cult 6(1):81–102
9. Bishop PA, Herron RL (2015) Use and misuse of the likert item responses and other ordinal measures. Int J Exerc Sci 8(3):297–302
10. Blight D, Davis D, Olsen A (1999) The internationalisation of higher education. In: Harry K (ed) Higher education through open and distance learning. Routledge, New York, pp 15–31
11. Blunch N (2012) Introduction to structural equation modeling using IBM SPSS statistics and AMOS. Sage, London
12. Darwish A, Lakhtaria KI (2011) The impact of the new Web 2.0 technologies in communication, development, and revolutions of societies. J Adv Inf Technol 2(4):204–216
13. Davis C (2009) Web 2.0 definition, usage, and self-efficacy: a study of graduate library school students and academic librarians at colleges and universities with ALA accredited degree programs, The University of Alabama
14. Diggins M (2004) Teaching and learning communication skills in social work education. Retrieved 2017, 22 Feb from http://www.scie.org.uk/publications/guides/guide05/
15. DiMaggio P, Hargittai E, Neuman R, Robinson J (2001) Social Implications of the Internet. Ann Rev Sociol 27:307–336
16. Duygu A, Zahide Y (2015) Using social networking sites for teaching and learning: students' involvement in and acceptance of facebook® as a course management system. J Educ Comput Res 52(2):155–179. https://doi.org/10.1177/0735633115571299
17. Dwyer C , Hiltz S, Passerini K (2007) Trust and privacy concern within social netowrking sites: a comparison of facebook and myspace. Paper presented at the thirteenth americas conference on information systems (AMCIS) Keystonee
18. Edosomwan S, Prakasan SK, Kouame D, Watson J, Seymour T (2011) The history of social media and its impact on business. J Appl Manag Entrep 16(3):79
19. Forrester Research (2010) Social networking in the enterprise: benefits and inhibitors. In: Consulting F (ed) A commissioned study conducted by Forrester Consulting on behalf of Cisco Systems. Forrester Research, USA, pp 1–17
20. Gliem, Joseph A, & Gliem, Rosemary R. (2003). Calculating, interpreting, and reporting Cronbach's alpha reliability coefficient for Likert-type scales
21. Goktalay SB, Ozdilek Z (2016) Social networking in higher education in Turkey: students' use and perceptions. In: Issa T, Isaias P, Kommers P (eds) Social networking and education. Springer, Cham, Switzerland, pp 167–187
22. Grosseck G (2009) To use or not to use web 2.0 in higher education? Procedia Soc Behav Sci 1(1):478–482
23. Gupta S, Seth A (2014) Web 2.0 tools in higher education. Trends Inf Manag 10(1):1–11

24. Hardie E, Tee M (2007) Excessive internet use: the role of personality, loneliness and social support networks in internet addiction. Aust J Emerg Technol Soc 5(1):34–47
25. Hartley J (2014) Some thoughts on Likert-type scales. Int J Clin Health Psychol 14(1):83–86
26. Havakhor T, Soror AA, Sabherwal R (2018) Diffusion of knowledge in social media networks: effects of reputation mechanisms and distribution of knowledge roles. Inf Syst J 28(1):104–141
27. Issa T, Isaias P, Kommers P (2016) Social networking and education model (SNEM). In: Issa T, Isaias P, Kommers P (eds) Social networking and education global perspectives. Springer, Cham, Switzerland, pp 323–345
28. Issa T, Kommers P (2013) Social networking for web-based communities. Int J Web Based Communities 9(1):5–24
29. Issa T (2013) Online survey: best practice. In: Isaias P, Nunes MB (eds) Information systems research and exploring social artifacts: approaches and methodologies. IGI Global, USA, pp 1–19
30. Jeong EJ, Kim DH (2011) Social Activities, Self-Efficacy, Game Attitudes, and Game Addiction. Cyberpsychol Behav Soc Netw 14(4):213–221
31. Kraut R, Patterson M, Lundmark V (1998) Internet paradox: a social technology that reduces social involvement and psychological well being. Am Psychol 53:1017–1031
32. Lederer K (2012) Pros and cons of social media in the classroom. Campus Technol 25(5):1–2
33. Lin C-P, Anol B (2008) Learning online social support: an investigation of network information technology based on UTAUT. CyberPsychol Behav 11(3):268–272
34. Liu D, Brown BB (2014) Self-disclosure on social networking sites, positive feedback, and social capital among Chinese college students. Comput Hum Behav 38(0):213–219. https://doi.org/10.1016/j.chb.2014.06.003
35. Malita L (2011) Social Media time management tools and tips. Procedia Comput Sci 3:747–753
36. Marasco A, Buonincontri P, van Niekerk M, Orlowski M, Okumus F (2018) Exploring the role of next-generation virtual technologies in destination marketing. J Destin Mark Manage 9:138–148
37. Martínez-Alemán AM, Wartman KL (2008) Online social networking on campus: Understanding what matters in student culture. Routledge, New York
38. Monks TG, Kawwa T (1971) Social psychological aspects of comprehensive education. Int Rev Educ 17(1):66–76. https://doi.org/10.1007/BF01421371
39. Novakovich J, Miah S, Shaw S (2017) Designing curriculum to shape professional social media skills and identity in virtual communities of practice. Comput Educ 104:65–90
40. Osborne C (2012). The pros and cons of social media classrooms http://www.zdnet.com/article/the-pros-and-cons-of-social-media-classrooms/
41. Pempek T, Yermolayeva Y, Calvert S (2000) College Students' social networking experieences on Facebook. J Appl Dev Psychol
42. Thatcher J, Loughry M, Lim J, Mcknight H (2007) Internet Anxiety: an empirical study of the effects of Personality, beliefts, and social support. Inf Manag 44:353–363
43. Thomson H (2008) What is Web 2.0 technology?. Retrieved 30 March 2018 from http://copyright.unimelb.edu.au/__data/assets/pdf_file/0011/1773830/wikisblogsweb2blue.pdf
44. Tyagi S (2012) Adoption of Web 2.0 technology in higher education: a case study of universities in National Capital Region, India. Int J Educ Dev Use Inf Commun Technol 8(2):28
45. Tyler T (2002) Is the Internet Changing Social Life? It seems the more things change, the more they stay the same. J Soc Issues 58(1):195–205
46. Verma MK, Verma NK (2015) Use of web 2.0 technology by the Indian Institutes of Technology (IITs) and Indian Institutes of Management (IIMs): a comparative survey. Int J Libr Inf Stud 5(5):98–106
47. Waters RD, Burnett E, Lamm A, Lucas J (2009) Engaging stakeholders through social networking: how nonprofit organizations are using Facebook. Publ Relat Rev 35(2):102–106

48. Weaver A, Morrison B (2008) Soc Netw Comput 41(2):97–100
49. Williams B, Onsman A, Brown T (2010) Exploratory factor analysis: a five-step guide for novices. Journal of Emergency Primary Health Care (JEPHC) 8(3):1–13
50. Yadav A, Patwardhan A. (2016) Use and impact of web 2.0 tools in higher education: a literature review. https://www.researchgate.net/publication/306118900_Use_and_Impact_of_Web_20_Tools_in_Higher_Education_A_Literature_Review
51. Zueger PM, Katz NL, Popovich NG (2014) Assessing outcomes and perceived benefits of a professional development seminar series. Am J Pharmaceut Educ 78(8):150

Index

X. Li, K.-C. Wong (eds.), *Natural Computing for Unsupervised Learning*, Unsupervised and Semi-Supervised Learning,
https://doi.org/10.1007/978-3-319-98566-4

B

W

Z

Printed in the United States
By Bookmasters